Name ______________________________ Class ______

Skills Worksheet

Directed Reading

Section: Categories of Biological Classification

Complete each statement by writing the correct term or phrase in the space provided.

1. The science of naming and classifying organisms is called

________________________ .

2. The Greek philosopher and naturalist Aristotle grouped plants according to their ________________________ similarities.

3. Linnaeus's two-word system for naming organisms is called

________________________ ________________________ .

4. The basic biological unit in the Linnaean system of biological classification is ________________________ .

5. A(n) ________________________ is a taxonomic category containing similar species.

6. The scientific name of the willow oak is ________________________

________________________ .

7. The common name of *Quercus rubra* is the ________________________

________________________ .

Complete each statement by underlining the correct term or phrase in the brackets.

8. The first word of a scientific name is the [species / genus].

9. Oak trees are placed in the [species / genus] *Quercus*.

10. People in Great Britain call [*Erithacus rubicula* / *Turdus migratorius*] a robin.

11. The correct abbreviation of the scientific name for modern humans is [*H. sapiens* / *h.s.*].

Study the following categories of classification. Determine the correct order of the categories from largest to smallest. Write the number of each category in the space provided.

______ **12.** phylum

______ **13.** class

______ **14.** species

______ **15.** family

______ **16.** order

______ **17.** kingdom

______ **18.** genus

______ **19.** domain

Read each question, and write your answer in the space provided.

20. How did biologists name a particular type of organism before the mid-1700s?

____

____

____

____

21. Explain how the genus and species name of an organism is properly written.

____

____

____

____

Name ______________________ Class ______________ Date ____________

Skills Worksheet

Directed Reading

Section: How Biologists Classify Organisms

In the space provided, write the letter of the term or phrase that best completes each statement or best answers each question.

______ **1.** According to the biological species concept, the members of the same species actually can
- **a.** interbreed.
- **b.** have radically different characters.
- **c.** have different scientific names.
- **d.** converge.

______ **2.** Modern biologists determine species by studying an organism's
- **a.** convergent evolution.
- **b.** hybrid offspring.
- **c.** features with respect to its evolutionary history.
- **d.** population.

______ **3.** Wolves and dogs are members of different species, but they
- **a.** can produce infertile offspring.
- **b.** are reproductively isolated.
- **c.** are unable to interbreed.
- **d.** can produce fertile offspring.

______ **4.** An example of an ancestral character of birds and mammals is
- **a.** feathers.
- **b.** backbones.
- **c.** mammary glands.
- **d.** fur.

______ **5.** How many species have been described so far?
- **a.** 1 million species
- **b.** 1.5 million species
- **c.** 5 million species
- **d.** 10 million species

______ **6.** The biological species concept cannot account for species that
- **a.** have hybrid offspring.
- **b.** have large numbers of offspring.
- **c.** reproduce sexually.
- **d.** reproduce asexually.

Read each question, and write your answer in the space provided.

7. What is reproductive isolation?

____

____

____

____

8. What are hybrids?

____

____

____

____

In the space provided, explain how the terms in each pair are related to each other.

9. convergent evolution, analogous characters

____

____

____

____

10. cladistics, derived traits

____

____

____

____

Read each question, and write your answer in the space provided.

11. What is phylogeny?

____

____

12. Why must biologists be able to distinguish homologous traits from analogous traits?

____

____

____

____

13. What is a cladogram?

____

____

____

____

In the space provided, write the letter of the term or phrase that best completes each statement or best answers each question.

______**14.** A great strength of cladograms is that they are

a. complicated.
b. simple.
c. biased.
d. objective.

______**15.** Where does evolutionary systematics place birds?

a. in the same class as reptiles
b. in the same genus as mammals
c. in their own class
d. in their own genus

______**16.** Evolutionary systematics allows biologists to classify organisms using all available evidence and
a. cladistics.
b. their own judgment.
c. their knowledge of cladistics.
d. their understanding of analogous features.

______**17.** In evolutionary systematics, evolutionary relationships are displayed in a branching diagram called a
a. helix.
b. ladder.
c. cladogram.
d. phylogenetic tree.

______**18.** A cladogram is based entirely on whether an organism has or does not have a(n)
a. derived character.
b. ancestral character.
c. analogous character.
d. homologous character.

Name ______________________ Class ______________ Date ____________

Skills Worksheet

Active Reading

Section: Categories of Biological Classification

Read the passage below. Then answer the questions that follow.

Linnaeus worked out a broad system of classification for plants and animals in which an organism's form and structure are the basis for arranging specimens in a collection. He later organized the genera and species that he described into a ranked system of groups that increase in inclusiveness. The different groups into which organisms are classified have expanded since Linnaeus's time and now consist of eight levels.

Similar genera are grouped into a family. Similar families are combined into an order. Orders with common properties are united in a class. Classes with similar characteristics are assigned to a phylum. Similar phyla are collected into a kingdom. Similar kingdoms are grouped into domains. All living things are grouped into one of three domains. Two domains, Archaea and Bacteria, are each composed of a single kingdom of prokaryotes. The third domain, Eukarya, contains all four kingdoms of eukaryotes.

SKILL: READING EFFECTIVELY

Read each question, and write your answer in the space provided.

1. What did Linnaeus use as the basis for classifying organisms in a collection?

____

____

2. The second sentence of this passage states that Linnaeus described a "ranked system of groups that increase in inclusiveness." What does this mean?

____

____

____

3. How many kingdoms exist in the modern system of classification? What are they?

____

____

SKILL: INTERPRETING GRAPHICS

4. The figure below shows the eight levels of the classification system. Using the information contained in the passage, insert the correct label in the space provided on the left side of the figure. On the right side of the figure, compose a sentence that describes the level. Use a separate sheet of paper if necessary. The first one has been done for you.

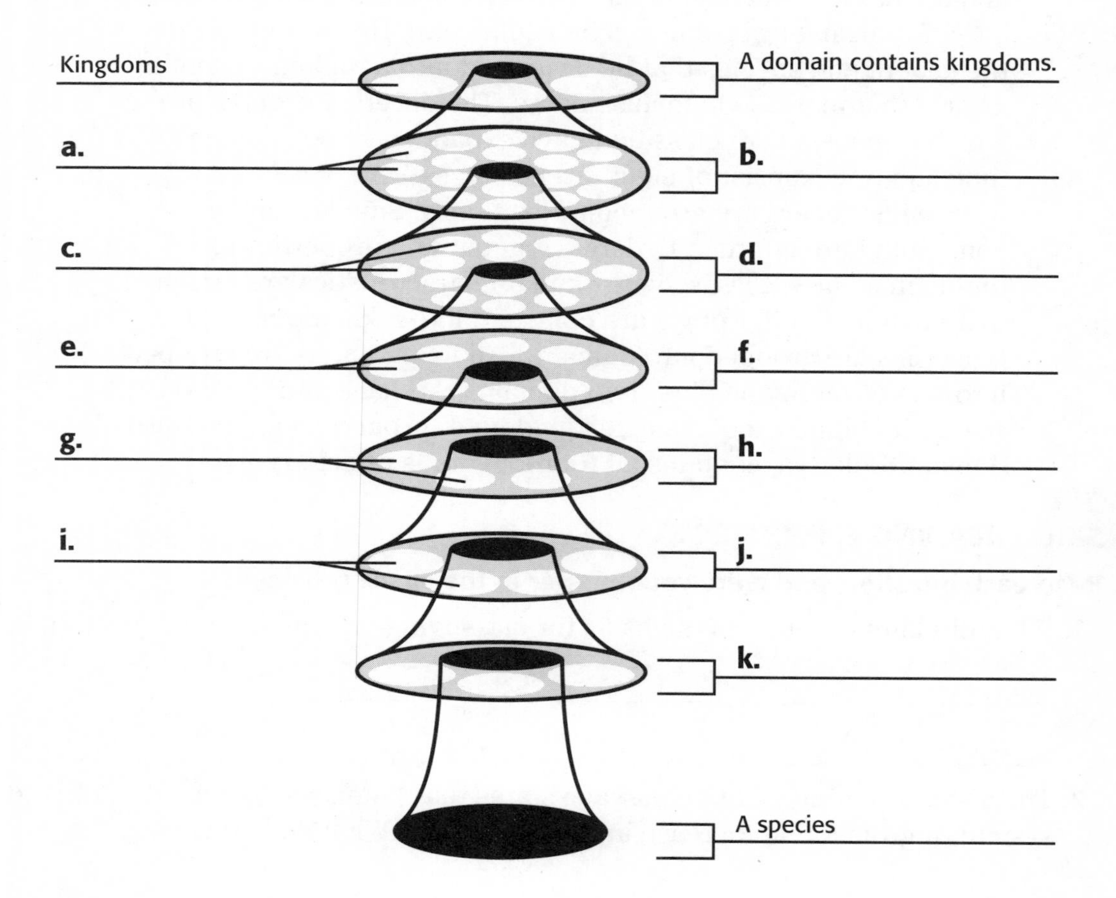

An analogy is a comparison. In the space provided, write the letter of the term or phrase that best completes the analogy.

______ **5.** Class is to order as order is to

a. kingdom.
b. species.
c. phylum.
d. families.

Name ______________________ Class ______________ Date ____________

Skills Worksheet

Active Reading

Section: How Biologists Classify Organisms

Read the passage below. Then answer the questions that follow.

Most biologists analyze evolutionary relationships using cladistics. **Cladistics** is a method of analysis that reconstructs phylogenies by inferring relationships based on shared characters. Cladistics can be used to hypothesize the sequence in which different groups of organisms evolved. To do this, cladistics focuses on the nature of the characters in different groups of organisms. With respect to two different groups, a character is defined as an **ancestral character** if it evolved in a common ancestor of both groups. Thus when considering the relationship between birds and mammals, a backbone is an ancestral character. Having feathers, however, is a derived character. A **derived character** evolved in an ancestor of one group but not of the other. Feathers evolved in an ancestor of birds that was not also ancestral to mammals.

SKILL: READING EFFECTIVELY

Read each question, and write your answer in the space provided.

1. How does cladistics reconstruct phylogenies?

____

____

2. What type of information is determined through cladistics?

____

____

3. How are derived characters and cladistics related?

____

____

____

____

In the space provided, write the letter of the term or phrase that best completes the statement.

______ **4.** Derived traits are a set of unique characteristics
- **a.** found in a single species.
- **b.** that mammals lack.
- **c.** found in a particular group of organisms.
- **d.** Both (a) and (b)

Name ______________________ Class ______________ Date ______________

Skills Worksheet

Vocabulary Review

Complete each statement by writing the correct term or phrase from the list below in the space provided.

analogous character	convergent evolution	kingdom
binomial nomenclature	derived characters	order
biological species	domain	phylogenetic tree
cladistics	evolutionary systematics	phylogeny
cladogram	family	phylum
class	genus	taxonomy

1. The classification level in which classes with similar characteristics are grouped is called a(n) ______________________ .

2. When taxonomists give varying subjective degrees of importance to characters, they are applying ______________________ ______________________ .

3. Reconstructing phylogenies by inferring relationships based on similarities derived from a common ancestor without considering the "strength" of a character is called ______________________ .

4. The evolutionary history of a species is its ______________________ .

5. Orders with common properties are combined into a(n) ______________________ .

6. Similar families are combined into a(n) ______________________ .

7. The classification level in which similar genera are grouped is called a(n) ______________________ .

8. A similar feature that evolved through convergent evolution is called a(n) ______________________ ______________________ .

9. In ______________________ ______________________ , organisms evolve similar features independently, often because they live in similar habitats.

10. A(n) ______________________ is a branching diagram used to show evolutionary relationships in groups of shared derived characters.

11. The most general level of classification is ______________________ .

Vocabulary Review *continued*

12. A(n) ______________________ is a taxonomic category containing similar species.

13. Linnaeus developed a system for naming and classifying organisms, which is called ______________________ .

14. A(n) ______________________ ______________________ is a group of interbreeding or potentially interbreeding natural populations that are reproductively isolated from other such groups.

15. Unique characteristics used in cladistics are called ______________________ ______________________ .

16. The two-word system for naming organisms is called ______________________ ______________________ .

17. A(n) ______________________ contains many phyla.

18. In evolutionary systematics, evolutionary relationships are displayed in a branching diagram called a ______________________ ______________________ .

Name ______________________ Class ____________ Date ____________

Skills Worksheet

Science Skills

Organizing Information

Use the list of animals below to complete items 1–3.

bat	frog	horse	rabbit
chicken	goldfish	octopus	spider
eagle	grasshopper	polar bear	whale

1. Organize the animals in the list above into the following four groups of classification. In the space provided, write the names of the animals from the list above that have each of the described characteristics. Some animals may fall into more than one group.

Group 1

a. Lives on land ____

____

b. Lives in water ____

____

Group 2

c. Catches live prey ____

____

d. Vegetarian ____

____

Group 3

e. Lays eggs ____

____

f. Live birth ____

____

Group 4

g. Lays eggs; has an internal skeleton ______________________

____

h. Lays eggs; has an internal skeleton; has feathers ______________________

____

i. Live birth; lives on land ______________________

____

j. Live birth; lives in water ______________________

____

Read each question, and write your answer in the space provided.

2. Which of the four groups above do you think classifies the animals in a manner that best demonstrates their evolutionary relationships? Explain.

____

____

____

____

____

____

____

____

3. Are there any animals that did not fit well into any of the categories listed in Groups 1–4? Explain.

____

____

____

____

Name ______________________ Class ______________ Date ______________

Skills Worksheet

Concept Mapping

Using the terms and phrases provided below, complete the concept map showing the use of taxonomy.

binomial nomenclature
biological classification
class
domain
genus
kingdom
order
phylogenies
phylum
species

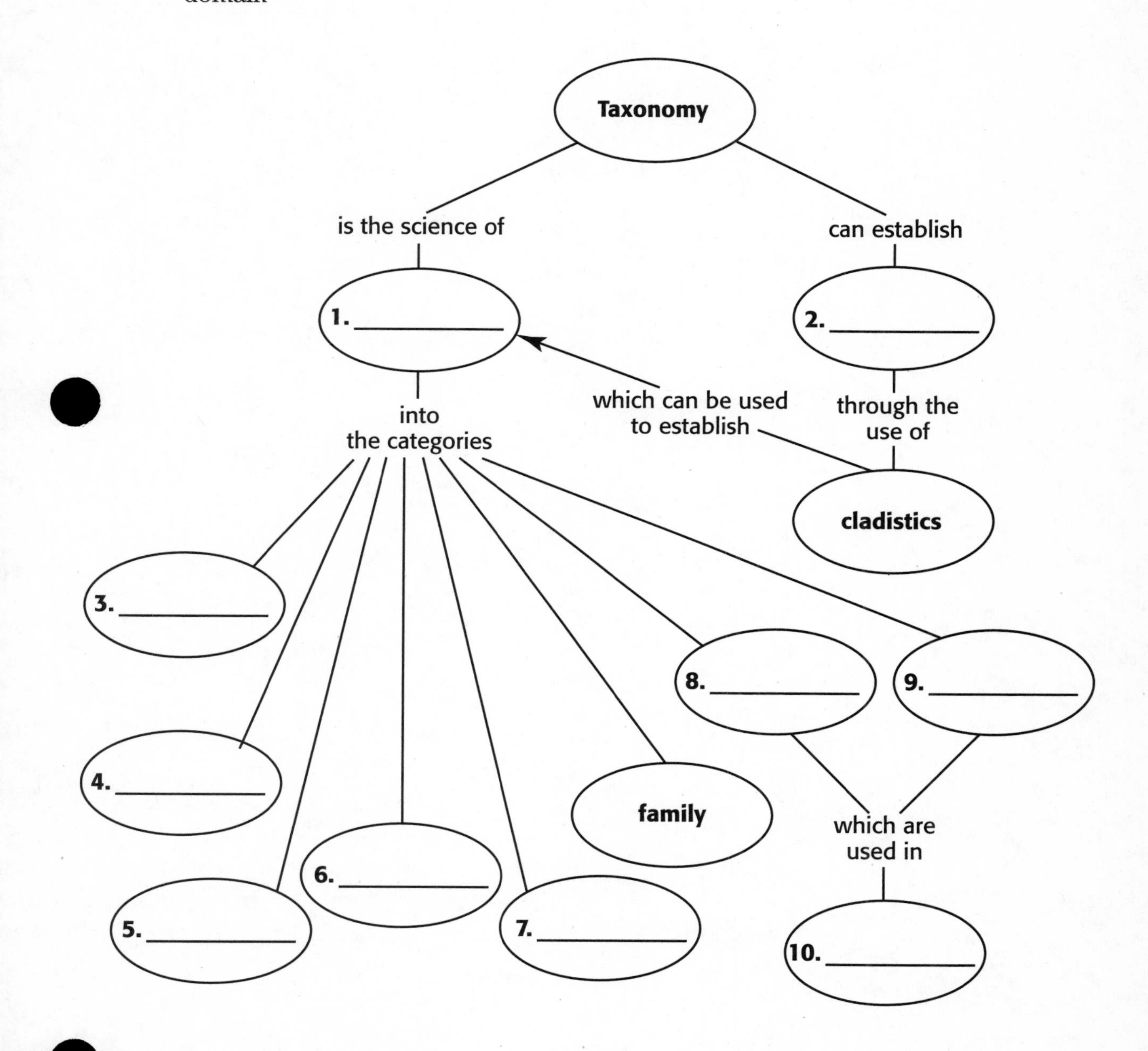

Name ______________________ Class ______________ Date ____________

Skills Worksheet

Critical Thinking

Work-Alikes

In the space provided, write the letter of the term or phrase that best describes how each numbered item functions.

______ **1.** scientific name

______ **2.** eight levels of classification

______ **3.** hybrid

______ **4.** cladogram

a. chocolate milk

b. a person's last name and first name

c. family tree

d. a child's toy block set in which small blocks fit into successively larger blocks

Cause and Effect

In the space provided, write the letter of the term or phrase that best matches each cause or effect given below.

Cause	**Effect**
5. ________________	Linnaeus devised binomial nomenclature
6. unique set of characteristics in a particular set of organisms	________________
7. ________________	convergent evolution
8. scientists have named only 500,000 of 6 billion species in the tropics	________________
9. subjective analysis of evolutionary relationships	________________

a. analogous characters

b. evolutionary systematics

c. polynomials were unwieldy and awkward

d. our knowledge of these organisms is limited

e. cladistics

Trade-offs

In the space provided, write the letter of the bad news item that best matches each numbered good news item below.

Good News

______ **10.** The biological species concept works for most members of the animal kingdom.

______ **11.** Reproductive isolation occurs when a barrier keeps organisms from breeding.

______ **12.** A cladogram is very objective.

Bad News

a. It does not take into account the strength of a character.

b. Sometimes reproductive barriers are not complete.

c. It fails to describe species that reproduce asexually.

Linkages

In the spaces provided, write the letters of the two terms or phrases that are related by the term or phrase in the middle. The choices can be placed in any order. Some choices may be used more than once.

13. ______ phylum ______

14. ______ family ______

15. ______ genera (genus) ______

16. ______ Apidae ______

17. ______ interbreeding ______

18. ______ cladogram ______

a. kingdom

b. *Apis mellifera*

c. derived characters

d. fertile hybrids

e. wolves and dogs

f. genus

g. Hymenoptera

h. species

i. evolutionary relationships among groups of organisms

j. class

k. order

l. family

Analogies

An analogy is a relationship between two pairs of terms or phrases written as a : b :: c : d. The symbol : is read as "is to," and the symbol :: is read as "as." In the space provided, write the letter of the pair of terms or phrases that best completes the analogy shown.

______ **19.** native language : common name ::
- **a.** British : scientific name
- **b.** Latin : scientific name
- **c.** English : common name
- **d.** Latin : common name

______ **20.** two : scientific name ::
- **a.** three : kingdom
- **b.** eight : biological classification
- **c.** two : family
- **d.** three : scientific name

Name ______________________ Class ______________ Date ______________

Skills Worksheet

Test Prep Pretest

In the space provided, write the letter of the term or phrase that best completes each statement or best answers each question.

______ **1.** In one of the earliest classification systems, Aristotle grouped plants and animals according to
a. basic categories.
b. structural similarities.
c. genus.
d. major characteristics.

______ **2.** Although Linnaeus used the Latin polynomial system in his books, he created his own
a. rules of grammar.
b. taxonomic categories.
c. evolutionary systematics.
d. two-word shorthand system, also in Latin.

______ **3.** Scientists classify organisms by studying their forms and
a. structures.
b. size.
c. method of reproduction.
d. cladograms.

______ **4.** Cladograms determine evolutionary relationships between organisms by examining
a. the strength of a character.
b. the degree of difference between organisms.
c. shared ancestral characters.
d. shared derived characters.

______ **5.** All members of the kingdom Animalia are multicellular
a. autotrophs whose cells have walls.
b. heterotrophs whose cells have walls.
c. heterotrophs whose cells lack walls.
d. autotrophs whose cells lack walls.

______ **6.** Biological species, as defined by Ernst Mayr,
a. are closely related.
b. are interbreeding natural populations.
c. produce infertile offspring.
d. produce infertile hybrids.

______ **7.** The characteristics that scientists use in cladistics are
a. analogous structures.
b. shared derived characters.
c. convergent structures.
d. shared homologous traits.

______ **8.** Bird wings and insect wings are
- **a.** homologous traits.
- **b.** derived traits.
- **c.** analogous traits.
- **d.** phylogenetic traits.

______ **9.** The biological species concept cannot be applied to
- **a.** species that can produce fertile hybrids.
- **b.** all bacteria.
- **c.** species that reproduce asexually.
- **d.** All of the above

______ **10.** Scientific names
- **a.** must have three Latin words and correct Latin grammar.
- **b.** include the genus and family.
- **c.** have rules established by British and American biologists.
- **d.** enable biologists to communicate regardless of their native language.

______ **11.** Which of the following lists the eight classification levels in proper descending order?
- **a.** domain, kingdom, phylum, class, order, family, genus, species
- **b.** kingdom, domain, phylum, order, class, family, genus, species
- **c.** kingdom, phylum, family, class, domain, order, genus, species
- **d.** phylum, kingdom, domain, class, order, family, genus, species

______ **12.** The scientific naming system requires all of the following EXCEPT that
- **a.** both words should be underlined or italicized.
- **b.** the genus is to be capitalized.
- **c.** the species should be the second word.
- **d.** the genus is never abbreviated.

Complete each statement by writing the correct term or phrase in the space provided.

13. The naming system developed by Linnaeus is called ______________________ ______________________.

14. One genus can include several ______________________.

15. Ernst Mayr developed the concept that a(n) ______________________ ______________________ is reproductively isolated from other groups.

16. When ______________________ ______________________ are incomplete, closely related species can produce hybrids.

17. The biological species concept works best for most members of the kingdom ______________________.

18. Similar features in organisms that do not share a recent common ancestor are called ________________ ________________ .

19. Scientists use evidence of ________________ characters to reconstruct evolutionary history.

20. The evolutionary history of a species is called its ________________ .

Read each question, and write your answer in the space provided.

21. Explain the difference between homologous characters and analogous characters. Give an example of each.

____

____

____

____

____

____

____

22. Which classification system would probably be used first if a scientist discovered five unknown plants? Explain.

____

____

____

23. Explain why Mayr's concept of biological species has limited applications.

____

____

____

____

Questions 24 and 25 refer to the figure below. The phylogenetic tree shown indicates the evolutionary relationships for a hypothetical group of modern organisms, labeled *1–5*, and their ancestors, labeled *A–E*.

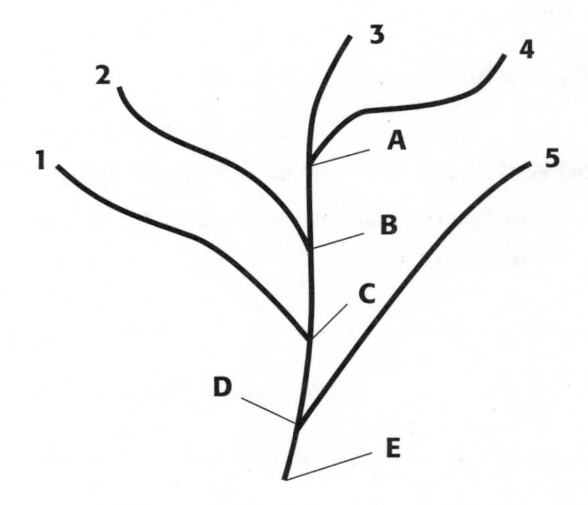

24. Which two modern organisms are likely to be most closely related?

____

25. What was the most recent common ancestor of the organisms labeled *1* and *5*?

____

Name ______________________ Class ______________ Date ______________

Assessment

Quiz

Section: Categories of Biological Classification

In the space provided, write the letter of the term or phrase that best completes each statement or best answers each question.

_____ **1.** The naturalist who developed the two-word naming system for organisms was
a. Aristotle.
b. Carolus Linnaeus.
c. Charles Darwin.
d. Gregor Mendel.

_____ **2.** The polynomial naming system in use until the mid-1700s was
a. a simpler system for naming organisms.
b. not very descriptive.
c. cumbersome and awkward.
d. more scientific than binomial nomenclature.

_____ **3.** Which of the following is the appropriate way to write a species name?
a. *Quercus rubra*
b. *Q. rubra* (after first use)
c. Quercus rubra
d. All of the above

_____ **4.** The full classification of the honeybee *Apis mellifera* describes all of the following characteristics EXCEPT
a. eukaryotic, multicellular heterotroph with no cell walls
b. three major body parts and three pairs of legs
c. one pair of wings and a stinger
d. jointed appendages and a hard outer skin of chitin

_____ **5.** The basic biological unit of the Linnaean system is
a. species.
b. genus.
c. phylum.
d. kingdom.

In the space provided, write the letter of the description that best matches the term or phrase.

_____ **6.** domain
_____ **7.** phylum
_____ **8.** order
_____ **9.** genus
_____ **10.** family

a. contains genera
b. contains families
c. contains kingdoms
d. contains classes
e. contains species

Name ______________________ Class ______________ Date ______________

Assessment

Quiz

Section: How Biologists Classify Organisms

In the space provided, write the letter of the term or phrase that best completes each statement or best answers each question.

______ **1.** Which of the following is used by biologists to determine if organisms are of the same species?
a. comparisons of appearance
b. comparisons of structure
c. comparisons of reproductive potential
d. All of the above

______ **2.** Sometimes individuals of different species interbreed and produce offspring called
a. hybrids.
b. analogues.
c. derived characters.
d. cladistics.

______ **3.** With respect to the biological species concept, within many groups of organisms, there are no
a. variations among individuals.
b. barriers to hybridization.
c. closely related species.
d. asexually reproducing species.

______ **4.** When similar characteristics in different organisms have evolved independently of one another, it is called
a. homologous evolution.
b. convergent evolution.
c. cladistics.
d. evolutionary systematics.

______ **5.** Which is an important difference between cladistic evolutionary analysis and evolutionary systematic analysis?
a. Cladistics analyzes evolutionary relationships.
b. Evolutionary systematics employs a branching diagram.
c. Cladistics is an objective analysis based on the principle of shared derived characters.
d. Evolutionary systematics is the more objective analysis.

In the space provided, write the letter of the description that best matches the term or phrase.

______ **6.** reproductive isolation

______ **7.** hybrid

______ **8.** asexually reproducing species

______ **9.** cladistics

______ **10.** evolutionary systematics

a. biological species concept fails

b. degree of difference not considered

c. barrier to hybridization

d. varying degrees of importance to characters

e. the result of the interbreeding of closely related species

Name ______________________ Class ______________ Date ______________

Assessment

Chapter Test

Classification of Organisms

In the space provided, write the letter of the term or phrase that best completes each statement or best answers each question.

______ **1.** The science of naming and classifying organisms is called
a. binomial nomenclature.
b. polynomial nomenclature.
c. taxonomy.
d. evolutionary systematics.

______ **2.** Linnaeus developed a new naming system because
a. he disagreed with the current classifications.
b. the polynomial system was too complicated.
c. the old one did not use Latin.
d. the polynomial system's descriptions were too brief.

______ **3.** Which of the following do biologists NOT use to classify organisms?
a. homologous structures
b. derived characters
c. appearance
d. analogous structures

______ **4.** An organism's scientific name consists of its
a. genus and species.
b. genus and family.
c. species and family.
d. common name and Latin name.

______ **5.** Biological species, as defined by Ernst Mayr, are
a. always closely related.
b. always reproductively isolated.
c. members of different genera.
d. None of the above

______ **6.** Honeybees, as members of the kingdom Animalia, are related to
a. wasps.
b. birds.
c. spiders.
d. All of the above

______ **7.** Linnaeus's classification system was based on which of the following characteristics?
a. form and structure
b. DNA
c. behavior
d. phylogenetic relationships

______ **8.** Convergent evolution leads to
a. shared homologous characters.
b. infertile hybrid offspring.
c. shared analogous characters.
d. fertile cladistic offspring.

_____ **9.** Classes with similar characteristics are grouped into the same
a. kingdom.
b. phylum.
c. species.
d. order.

_____ **10.** Analogous characters are derived from
a. a recent common ancestor.
b. inferred relationships.
c. a distant common ancestor.
d. independent sources.

_____ **11.** Biologists use cladograms to
a. evaluate the importance of characters.
b. estimate the degree of difference between organisms.
c. hypothesize the sequence in which different groups evolved.
d. analyze evolutionary relationships subjectively.

_____ **12.** The biological species concept fails to describe species that
a. reproduce asexually.
b. reproduce sexually.
c. are members of the kingdom Animalia.
d. None of the above

Questions 13 and 14 refer to the table below.

Classification of Three Different Organisms

Organism	Class	A	Family	Genus
Bacterium	Scotobacteria	Spirochaetales	Spirochaetaceae	*Cristispira*
Box elder	Dicotyledones	Sapindales	Aceraceae	*Acer*
Human	Mammalia	Primates	Hominidae	B

_____ **13.** Which level of classification is represented by the box labeled *A*?
a. kingdom
b. phylum
c. division
d. order

_____ **14.** Which of the following best fits the box labeled *B*?
a. *sapiens*
b. *Canis*
c. *Homo*
d. Animalia

In the space provided, write the letter of the description that best matches the term or phrase.

______ **15.** taxonomy

______ **16.** binomial nomenclature

______ **17.** domain

______ **18.** phylogeny

______ **19.** convergent evolution

______ **20.** ancestral character

a. the most inclusive classification of organisms

b. an organism's evolutionary history

c. similarities that evolved in a common ancestor of two different groups

d. the science of naming and classifying organisms

e. two-word system for naming organisms

f. similarities that evolve in organisms that are not closely related to one another

Name ______________________ Class ______________ Date ______________

Assessment

Chapter Test

Classification of Organisms

In the space provided, write the letter of the term or phrase that best completes each statement or best answers each question.

______ **1.** The Greek philosopher and naturalist who grouped plants and animals according to their structural similarities more than 2,000 years ago was
a. Mayr.
b. Linnaeus.
c. Aristotle.
d. Darwin.

______ **2.** Until the mid-1700s, organisms were rarely known to everyone by the same name because
a. species varied from place to place.
b. polynomials were often changed by biologists.
c. not very many people spoke or read Latin.
d. all species were not named yet.

______ **3.** An organism once named *Apis pubescens, thorace subgriseo, abdomine fusco, pedibus posticis glabis, untrinque margine ciliatus* is the
a. red oak.
b. willow oak.
c. Africanized honeybee.
d. European honeybee.

______ **4.** When can the genus name of an organism be abbreviated?
a. Always, if it is common and everyone is familiar with the species.
b. Never, because it could be confused with another organism.
c. Sometimes, after it's first use is spelled out.
d. Occasionally, depending on the amount of space there is to write.

______ **5.** Who established the rules for naming a species?
a. the scientist who discovered the species
b. an international commission of scientists
c. the government of the country where the species was discovered
d. Carolus Linnaeus

______ **6.** Linnaeus was the first to organize the genera and species that he described into
a. a ranked system of groups that increase in inclusiveness.
b. a random collection of groups.
c. several collections of major kinds of organisms, such as horses, fishes, and insects.
d. a ranked system of groups that increase in exclusiveness.

______ **7.** The domain Eukarya is composed of
a. all prokaryotes.
b. all archaebacteria.
c. all animals.
d. all eukaryotes.

______ **8.** In nature, reproductive barriers between sexually reproducing species are not always
a. effective.
b. complete.
c. productive.
d. natural.

______ **9.** When two separate species interbreed and produce fertile offspring, the two species are
a. distantly related.
b. entirely unrelated.
c. closely related.
d. not related.

______ **10.** A biological species is a group of natural populations that are interbreeding or could interbreed, and are
a. fertile hybrids.
b. the result of convergent evolution.
c. very different in appearance and structure.
d. reproductively isolated from other such groups.

______ **11.** *Canis lupus* and *Canis familiaris* produce
a. *Canis lupus* offspring.
b. fertile hybrid offspring.
c. *Canis familiaris* offspring.
d. infertile hybrid offspring.

______ **12.** In the tropics, what percentage of species has been named?
a. 100%
b. 50–100%
c. 20–30%
d. 5–10%

______ **13.** In which way are the wings of birds different from the wings of insects?
a. They do not perform the same function.
b. They evolved independently of each other.
c. They are structurally different.
d. Both (a) and (b)

______ **14.** In a whale and a dog, limbs are
- **a.** a derived character.
- **b.** an ancestral character.
- **c.** an analogous character.
- **d.** None of the above

______ **15.** A phylogenetic tree is
- **a.** based on derived characters without consideration for the "strength" of the characters.
- **b.** an objective analysis of evolutionary relationships.
- **c.** a diagram of evolutionary relationships through evolutionary systematics.
- **d.** a diagram of evolutionary relationships through cladistics.

In the space provided, write the letter of the description that best matches the term or phrase.

______ **16.** domain

______ **17.** phylogeny

______ **18.** convergent evolution

______ **19.** family

______ **20.** cladogram

______ **21.** species

______ **22.** binomial nomenclature

a. objective evolutionary analysis

b. contains genera

c. least inclusive

d. two-word naming system

e. leads to analogous characters

f. most inclusive

g. evolutionary history

Read each question, and write your answer in the space provided.

23. Describe the advantages of the modern system of binomial nomenclature.

____

____

____

____

____

____

24. In a zoo research facility, biologists are working to save an endangered species of prairie bird. In the zoo environment, the bird and another similar bird produced fertile offspring. The two birds never breed in nature, however, because they reproduce at different times of the year. Are the birds the same species? Explain why or why not.

____

____

____

____

____

____

25. Explain how convergent evolution leads to analogous characters. Explain your answer in terms of natural selection.

____

____

____

____

____

____

Name ______________________ Class ______________ Date ______________

Quick Lab

DATASHEET FOR IN-TEXT LAB

Using a Field Guide

You can use a standard pictoral field guide or a dichotomous key to help you identify species of plants, animals, or other organisms.

MATERIALS

- paper and pencil
- a plant or an animal field guide

Procedure

1. **CAUTION: Wear protective gloves when handling any wild plant. Keep your hands away from your face.** Using a dichotomous key or other field guide, identify several species of plants that share the same phylum and class. Collect specimens only if your teacher tells you to do so.
2. Try to identify two plants of the same genus.
3. Record the scientific name of each specimen in the table below.

Specimen	Genus name	Binomial species name	Identifying characteristics
A			
B			
C			

4. Read the description of each species in the field guide. Determine the set of characteristics that fit each specimen. Write these characteristics in the table.

Analysis

1. **List** the characteristics shared by two specimens that are in the same genus but are different species.

____

____

2. **Describe** how the binomial names of these two species show that they are members of the same genus.

____

____

3. **Identify** the key characteristics your field guide uses to tell these two species apart.

____

____

4. **Critical Thinking**
Analyzing Data Based on your observations, are two species from the same genus more similar or less similar than two species from different genera?

____

____

Name ______________________ Class ______________ Date ____________

Data Lab

DATASHEET FOR IN-TEXT LAB

Analyzing Taxonomy of Mythical Organisms

Background

Classification of organisms often requires grouping organisms based on their characteristics. Use the following list of mythological organisms and their characteristics to complete the analysis.

- **Pegasus** stands 6 ft tall, has a horse's body, a horse's head, four legs, and two wings.
- **Centaur** stands 6 ft tall, has a horse's body with a human torso, a male human head, and four legs.
- **Griffin** stands 4–6 ft tall, has a lion's body, an eagle's head, four legs, two wings, fur on its body, and feathers on its head and wings.
- **Dragon** can grow to several hundred feet, has a snake-like body, from 1 to 3 reptile-like heads, four legs, scales, and breathes fire.
- **Chimera** stands 6 ft tall, has a goat's body, snake's tail, four legs, a lion's head, fur on its body and head, scales on its tail, and breathes fire.
- **Hydra** is several hundred feet long, has a long body with four legs and a spiked tail, 100 snake heads, scales, and is poisonous.

Analysis

1. **Identify** the characteristics that you think are the most useful for grouping the organisms into separate groups.

____

____

2. **Classify** the organisms into at least three groups based on the characteristics you think are most important.

____

____

____

3. **Evaluate** the use of the biological species concept to classify these mythical organisms.

____

____

____

Name ______________________ Class ______________ Date ______________

Data Lab

DATASHEET FOR IN-TEXT LAB

Making a Cladogram

Background

A cladogram is a model that represents a hypothesis about the order in which organisms evolved from a common ancestor. Scientists construct a cladogram by first analyzing characters in a data table. The absence of a vascular system and the absence of seeds is ancestral. Use the data below to construct a cladogram on a separate sheet of paper.

Characters		
Plants	**Seeds**	**Vascular system**
Horsetails	No	Yes
Liverworts	No	No
Pine trees	Yes	Yes

Analysis

1. **Identify** the out-group.

____

____

2. **Name** the least common derived character.

____

____

3. **List** the order in which the plants in the table would be placed on a cladogram.

____

____

Name ______________________ Class ______________ Date ____________

Skills Practice Lab

DATASHEET FOR IN-TEXT LAB

Making a Dichotomous Key

SKILLS

- Identifying and comparing
- Organizing data

OBJECTIVES

- **Identify** objects using dichotomous keys.
- **Design** a dichotomous key for a group of objects.

MATERIALS

- 6 to 10 objects found in the classroom (e.g., shoes, books, writing instruments)
- stick-on labels
- pencil

Before You Begin

One way to identify an unknown organism is to use an **identification key,** which contains the major characteristics of groups of organisms. A **dichotomous key** is an identification key that contains pairs of contrasting descriptions. After each description, a key either directs the user to another pair of descriptions or identifies an object. In this lab, you will design and use a dichotomous key. A dichotomous key can be written for any group of objects.

1. Write a definition for each boldface term in the paragraph above.

____

____

____

____

2. Based on the objectives for this lab, write a question you would like to explore about making or using a dichotomous key.

____

____

Procedure

PART A: USING A DICHOTOMOUS KEY

1. Use the **Key to Forest Trees** to identify the tree that produced each of the leaves shown. Identify one leaf at a time. Always start with the first pair of statements (**1***a* and **1***b*). Follow the direction beside the statement that describes the leaf. Proceed through the key until you get to the name of a tree.

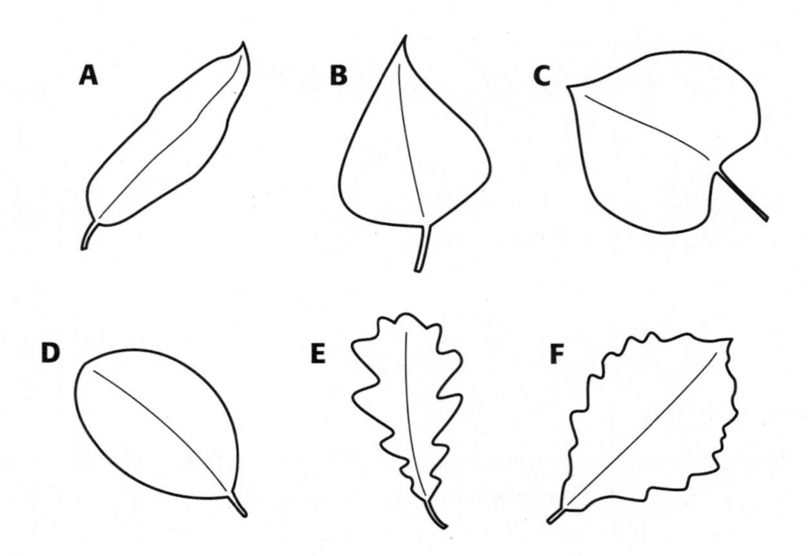

Key to Forest Trees		
1*a*	Leaf edge has no teeth, waves, or lobes	go to **2**
1*b*	Leaf edge has teeth, waves, or lobes	go to **3**
2*a*	Leaf has a bristle at its tip	**shingle oak**
2*b*	Leaf has no bristle at its tip	go to **4**
3*a*	Leaf edge is toothed	**Lombardy poplar**
3*b*	Leaf edge has waves or lobes	go to **5**
4*a*	Leaf is heart-shaped	**red bud**
4*b*	Leaf is not heart-shaped	**live oak**
5*a*	Leaf edge has lobes	**English oak**
5*b*	Leaf edge has waves	**chestnut oak**

PART B: DESIGN A DICHOTOMOUS KEY

2. Work with the members of your lab group to design a dichotomous key using the materials listed for this lab.

You Choose

As you design your key, decide the following:

a. what question you will explore

b. what objects your key will identify

c. how you will label personal property

d. what distinguishing characteristics the objects have

e. which characteristics to use in your key

f. how you will organize the data you will need for writing your key

3. Before you begin writing your key, have your teacher approve the objects your group has decided to work with.

4. Using the **Key to Forest Trees** as a guide, write a key for the objects your group selected. Remember, a dichotomous key includes pairs of contrasting descriptions.

5. Use your key to explore one of the questions written for step 2 of **Before You Begin.**

____

____

6. After each group has completed step 5, exchange keys and the objects they identify with another group. Use the key you receive to identify the objects. If the key does not work, return it to the group so corrections can be made.

PART C: CLEANUP

7. Clean up your work area and all lab equipment. Return lab equipment to its proper place. Wash your hands thoroughly before you leave the lab and after you finish all work.

Analyze and Conclude

1. Drawing Conclusions What tree produced each of the leaves shown in this lab?

____

____

____

____

2. Forming Hypotheses What other characteristics might be used to identify leaves using a dichotomous key?

____

____

____

____

3. Analyzing Methods How was the key your group designed dichotomous?

____

____

____

4. Evaluating Results Were you able to use another group's key to identify the objects for which it was written? If not, describe the problems you encountered.

____

____

____

____

____

5. Analyzing Methods Does a dichotomous key begin with general descriptions and then proceed to more specific descriptions or vice versa? Explain your answer, giving an example from your key.

____

____

____

6. Further Inquiry Write a new question about making or using keys that could be explored with another investigation.

____

____

____

Name ______________________ Class ______________ Date ____________

Exploration Lab

OBSERVATION

Classification

Taxonomy is the science of identifying and classifying living things. In taxonomy, organisms are grouped based on their similarities. By classifying organisms according to certain traits, biologists are able to determine which organisms may be related. As new technology arises, scientists can reevaluate the way organisms have been grouped. For example, DNA sequencing has given biologists an important characteristic with which to group organisms.

Biologists think there may be as many as 40 million different species of organisms on Earth. A *species* is a group of organisms capable of mating with one another to produce fertile offspring. Only about 1.5 million of the species on Earth have been named and described. Even though each species is different from all others, there are many similar characteristics among species. These similarities have made it possible for biologists to classify species into larger groups.

The naming of species based on their classification is very important because it makes communication about organisms more efficient. Biologists use the *Linnaean binomial system* to name a species. Each species is identified by two Latin names, or a *binomial*. The first name of the binomial identifies the *genus*, or group of species more closely related to one another than to any other group. The second name identifies a particular species within the genus. For example, grizzly bears belong to the species *Ursus arctos*. Polar bears are called *Ursus maritimus*. Both belong to the genus *Ursus* but are separate species.

In this lab, you will classify and give species names to imaginary organisms according to their characteristics. You will also create a dichotomous key that can be used to identify the organisms.

OBJECTIVES

Classify imaginary organisms according to their characteristics.

Create species names for the organisms.

Develop a dichotomous key for identifying the organisms.

MATERIALS

- pen or pencil

Procedure

1. The year is 2500. All bodies of water on Earth have become extremely polluted. Because of the pollution, almost every species of animal, plant, fungus, protist, bacteria, and virus that existed in the early 2000s has become extinct. However, because of natural selection, genetic engineering, and selective breeding programs, particular areas of the planet have been repopulated. The organisms now present are very different variations of the organisms that once existed.

The ten major species of organisms include the following:

A. chemosynthetic aquatic organisms that live on water contaminants

B. heterotrophic, aquatic organisms who work on "aqua farms"

C. terrestrial organisms with green, photosynthetic hair

D. anaerobic, heterotrophic, space travelers who must wear deoxygenated suits to survive on Earth

E. photosynthetic, legless space travelers who visit Earth every month and who have armlike tentacles

F. green-skinned, photosynthetic, transparent-haired terrestrial organisms

G. parasitic winged organisms that feed off of any terrestrial life form

H. decomposers that feed on pollution and solid waste

I. heterotrophic, microscopic decomposers that feed on aquaculture crops or any photosynthetic organism

J. heterotrophic aquatic crop, grown on the aqua farm, that feeds on bacteria or anything autotrophic

2. You have just arrived from another planet to classify these ten species. Create a classification scheme for the species to help you and others study them. Use the taxonomic tiers *kingdom, phylum, and genus.* First, group the species into two kingdoms, based on the organisms' characteristics. Record the names of your kingdoms in the two boxes on the next page. Also record the letters of the species (A–J) that belong to each kingdom.

3. On the next page draw the number of boxes that you will need for your scheme. Then record in the boxes the names of your phyla and the letters of the species that belong to each phylum. Use a vertical line to connect each phylum to the kingdom to which it belongs.

Kingdom

Phylum

Genus

4. Repeat step 3 for the genera in your classification scheme, except connect each genus to the phylum to which it belongs.
5. Create a species name for each of the ten species, using the Linnaean binomial system. Be sure to capitalize the genus name only. Write these names in the table on the next page.

TABLE 1 SPECIES LIVING IN THE YEAR 2500

Species description	Species name
A. chemosynthetic aquatic organisms	
B. heterotrophic aquatic farmer	
C. photosynthetic-haired terrestrial organisms	
D. anaerobic space travelers	
E. photosynthetic legless space travelers	
F. photosynthetic green-skinned terrestrial organisms	
G. parasitic winged organisms	
H. pollution and waste decomposers	
I. microscopic decomposers	
J. heterotrophic aquatic crop	

6. On the following page, create a dichotomous key for identifying the ten species. The key will need to be understood by your alien colleagues who will be identifying the species on their summer visits to Earth. To create a key, write two opposing descriptive statements that will divide the species into two groups. Start with a pair of statements that will separate the ten species into the two kingdoms you devised. At the end of each statement, write the number of the next set of opposing statements that will further describe and divide the species in each group. If the description of a species is complete, write the name of the species. Continue to write opposing statements until all the species are identified. For example:

Dichotomous Key		
1.	Organism is autotrophic.	Go to 2.
	Organism is not autotrophic.	Go to 3.
2.	Organism contains chlorophyll.	Go to 6.
	Organism does not contain chlorophyll.	*Name of species*

Name ______________________ Class ______________ Date ____________

DICHOTOMOUS KEY FOR SPECIES LIVING IN THE YEAR 2500

7. Exchange your completed key with another student. Using the new key, identify that student's names for the ten species.

Analysis

1. **Classifying** What characteristic did you use to group the species into two kingdoms?

____

____

____

____

2. **Describing Events** Were you able to follow the dichotomous key created by the other student? What problems, if any, did you have with the key you tested?

____

____

____

3. **Analyzing Data** Compare your key with the keys developed by other students. How are they similar? How are they different?

____

____

____

Conclusions

1. **Drawing Conclusions** Can more than one species occupy the same final place on your key? Why?

____

____

2. **Evaluating Methods** What advantage does the dichotomous key you developed have over descriptions of the ten species?

____

____

____

Extensions

1. **Building Models** Link the ten species listed in **Table 1** in two possible food chains, beginning with a producer (autotroph) and ending with a decomposer. Each food chain should have a minimum of four species. The same single species may appear in both food chains.
2. **Research and Communications** With a small group of your classmates, discuss the environmental issue of how pollution affects species and their evolution. How might genetic engineering or selective breeding programs change the course of the evolution of a species?

Name ______________________ Class ______________ Date ____________

Skills Practice Lab

OBSERVATION

Analyzing Amino-Acid Sequences

The biochemical comparison of proteins is a technique used to determine evolutionary relationships among organisms. Proteins consist of chains of amino acids. The sequence, or order, of the amino acids in a protein determines the type and nature of the protein. In turn, the sequence of amino acids in a protein is determined by the sequence of nucleotides in a gene. A change in the DNA nucleotide sequence (mutation) of a gene that codes for a protein may result in a change in the amino-acid sequence of the protein.

Biochemical evidence of evolution compares favorably with structural evidence of evolution. Even organisms that appear to have few physical similarities may have similar sequences of amino acids in their proteins and be closely related through evolution. Many researchers believe that the greater the similarity in the amino-acid sequences of two organisms, the more closely related they are in an evolutionary sense. Conversely, the greater the time that organisms have been diverging from a common ancestor, the greater the differences that can be expected in the amino-acid sequences of their proteins.

Two proteins are commonly studied in attempting to deduce evolutionary relationships from differences in amino-acid sequences. One is cytochrome *c*, and the other is hemoglobin. *Cytochrome* c is a protein used in cellular respiration and found in the mitochondria of many organisms. *Hemoglobin* is the oxygen-carrying molecule found in red blood cells.

In this lab, you will compare portions of human cytochrome c and hemoglobin molecules with the same portions of those molecules in other vertebrates. You will determine the differences in the amino-acid sequences of the molecules and deduce the evolutionary relationships among the vertebrates.

OBJECTIVES

Identify differences in the amino-acid sequences of the cytochrome *c* and hemoglobin molecules of several vertebrates.

Infer the evolutionary relationships among several vertebrates by comparing the amino-acid sequences of the same protein in those vertebrates.

MATERIALS

- scissors
- photocopy of cytochrome *c* amino-acid sequences
- photocopy of hemoglobin amino-acid sequences

Procedure

PART 1: CYTOCHROME *C*

A cytochrome c molecule consists of a chain of 104 amino acids. **Figure 1** shows the amino-acid sequences in a section of the cytochrome *c* molecules of eight vertebrates. The numbers refer to the positions of these amino acids.

1. Using a photocopy of the chart in **Figure 1,** cut out each vertebrate's amino-acid sequence. You should have eight amino-acid sequence strips.

FIGURE 1 CYTOCHROME *C* AMINO-ACID SEQUENCES

AA#	Horse	Chicken	Frog	Human	Shark	Turtle	Monkey	Rabbit
42	GLN	GLN	GLN	GLN	GLN	GLN	GLN	GLN
43	ALA	ALA	ALA	ALA	ALA	ALA	ALA	ALA
44	PRO	GLU	ALA	PRO	GLN	GLU	PRO	VAL
46	PHE	PHE	PHE	TYR	PHE	PHE	TYR	PHE
47	THR	SER	SER	SER	SER	SER	SER	SER
49	THR	THR	THR	THR	THR	THR	THR	THR
50	ASP	ASP	ASP	ALA	ASP	GLU	ALA	ASP
53	LYS	LYS	LYS	LYS	LYS	LYS	LYS	LYS
54	ASN	ASN	ASN	ASN	ASN	ASN	ASN	ASN
55	LYS	LYS	LYS	LYS	LYS	LYS	LYS	LYS
56	GLY	GLY	GLY	GLY	GLY	GLY	GLY	GLY
57	ILE	ILE	ILE	ILE	ILE	ILE	ILE	ILE
58	THR	THR	THR	ILE	THR	THR	ILE	THR
60	LYS	GLY	GLY	GLY	GLN	GLY	GLY	GLY
61	GLU	GLU	GLU	GLU	GLN	GLU	GLU	GLU
62	GLU	ASP	ASP	ASP	GLU	GLU	ASP	ASP
63	THR	THR	THR	THR	THR	THR	THR	THR
64	LEU	LEU	LEU	LEU	LEU	LEU	LEU	LEU
65	MET	MET	MET	MET	ARG	MET	MET	MET
66	GLU	GLU	GLU	GLU	ILE	GLU	GLU	GLU
100	LYS	ASP	SER	LYS	LYS	ASP	LYS	LYS
101	ALA	ALA	ALA	ALA	THR	ALA	ALA	ALA
102	THR	THR	CYS	THR	ALA	THR	THR	THR
103	ASN	SER	SER	ASN	ALA	SER	ASN	ASN
104	GLU	LYS	LYS	GLU	SER	LYS	GLU	GLU

2. Compare the amino-acid sequence of human cytochrome *c* with that of each of the other seven vertebrates by aligning the appropriate strips side by side.
3. For each nonhuman vertebrate's sequence, count the number of amino acids that differ from those in the human sequence. Write the number of differences next to that vertebrate's name on the strip.
4. When you have completed your comparisons, transfer your data to **Table 1.** As you do, list the seven nonhuman vertebrates in order from fewest differences to most differences.

TABLE 1 CYTOCHROME C AMINO-ACID SEQUENCE DIFFERENCES

Vertebrate	Number of differences from human cytochrome c

PART 2: HEMOGLOBIN

5. **Figure 2** shows the amino-acid sequences in corresponding parts of the hemoglobin molecules of five vertebrates. The parts of the chains shown are from amino acid number 87 to amino acid number 116, within a total sequence of 146 amino acids. Using a photocopy of the chart in **Figure 1,** cut out each vertebrate's amino-acid sequence to form five amino-acid sequence strips.

FIGURE 2 HEMOGLOBIN PROTEIN AMINO-ACID SEQUENCES

AA#	Human	Chimpanzee	Gorilla	Monkey	Horse
87	THR	THR	THR	GLN	ALA
88	LEU	LEU	LEU	LEU	LEU
89	SER	SER	SER	SER	SER
90	GLU	GLU	GLU	GLU	GLU
91	LEU	LEU	LEU	LEU	LEU
92	HIS	HIS	HIS	HIS	HIS
93	CYS	CYS	CYS	CYS	CYS
94	ASP	ASP	ASP	ASP	ASP
95	LYS	LYS	LYS	LYS	LYS
96	LEU	LEU	LEU	LEU	LEU
97	HIS	HIS	HIS	HIS	HIS
98	VAL	VAL	VAL	VAL	VAL
99	ASP	ASP	ASP	ASP	ASP
100	PRO	PRO	PRO	PRO	PRO
101	GLU	GLU	GLU	GLU	GLU
102	ASN	ASN	ASN	ASN	ASN
103	PHE	PHE	PHE	PHE	PHE
104	ARG	ARG	LYS	LYS	ARG
105	LEU	LEU	LEU	LEU	LEU
106	LEU	LEU	LEU	LEU	LEU
107	GLY	GLY	GLY	GLY	GLY
108	ASN	ASN	ASN	ASN	ASN
109	VAL	VAL	VAL	VAL	VAL
110	LEU	LEU	LEU	LEU	LEU
111	VAL	VAL	VAL	VAL	ALA
112	CYS	CYS	CYS	CYS	LEU
113	VAL	VAL	VAL	VAL	VAL
114	LEU	LEU	LEU	LEU	VAL
115	ALA	ALA	ALA	ALA	ALA
116	HIS	HIS	HIS	HIS	ARG

6. Use the strips you cut out to compare the amino-acid sequence of human hemoglobin with that of each of the other four vertebrates.
7. For each nonhuman vertebrate's sequence, count the number of amino acids that differ from the human sequence.
8. Write the number of differences next to that vertebrate's name on the strip.
9. When you have completed your comparisons, transfer your data to **Table 2.** Be sure to list the four vertebrates in order from fewest differences to most differences.

TABLE 2 HEMOGLOBIN AMINO-ACID SEQUENCE DIFFERENCES

Vertebrate	Number of differences from human hemoglobin

Analysis

1. **Identifying Relationships** According to the data in **Table 1,** which vertebrate is most closely related to humans? Which is least closely related to humans?

____

____

____

2. **Identifying Relationships** According to the data in **Table 2,** which vertebrate is most closely related to humans? Least closely related?

____

____

3. **Identifying Relationships** If the amino-acid sequences in gorillas and humans are similar, are the nucleotide sequences of their DNA also similar? Why?

____

____

Conclusions

1. **Evaluating Methods** Can you deduce from the data in **Table 1** that the chicken and the horse are closely related to each other? Why or why not?

____

____

2. **Drawing Conclusions** According to the data listed in **Table 2,** what conclusion can you make about how closely the three primates—chimpanzee, gorilla, and monkey—are related to each other?

____

____

3. **Applying Conclusions** In what way do proteins behave like molecular clocks?

____

____

____

____

Extensions

1. **Building Models** Use your data in **Table 1** to make a cladogram that shows the evolutionary relationships between humans and the seven vertebrates listed in the table.
2. **Research and Communications** Research how biologists determine the amino-acid sequence of a protein molecule.

Name ______________________ Class ______________ Date ______________

Quick Lab

DATASHEET FOR IN-TEXT LAB

Using a Field Guide

You can use a standard pictoral field guide or a dichotomous key to help you identify species of plants, animals, or other organisms.

MATERIALS

- paper and pencil
- a plant or an animal field guide

Procedure

1. **CAUTION: Wear protective gloves when handling any wild plant. Keep your hands away from your face.** Using a dichotomous key or other field guide, identify several species of plants that share the same phylum and class. Collect specimens only if your teacher tells you to do so.
2. Try to identify two plants of the same genus.
3. Record the scientific name of each specimen in the table below.

Specimen	Genus name	Binomial species name	Identifying characteristics
A			
B			
C			

4. Read the description of each species in the field guide. Determine the set of characteristics that fit each specimen. Write these characteristics in the table.

Analysis

1. **List** the characteristics shared by two specimens that are in the same genus but are different species.

   Answers will vary but may include leaf shape.

   ____

2. **Describe** how the binomial names of these two species show that they are members of the same genus.

   They have the same genus name.

   ____

3. **Identify** the key characteristics your field guide uses to tell these two species apart.

   Answers will vary

4. **Critical Thinking**
   Analyzing Data Based on your observations, are two species from the same genus more similar or less similar than two species from different genera?

   more similar

Name ______________________ Class ______________ Date ______________

Data Lab

DATASHEET FOR IN-TEXT LAB

Analyzing Taxonomy of Mythical Organisms

Background

Classification of organisms often requires grouping organisms based on their characteristics. Use the following list of mythological organisms and their characteristics to complete the analysis.

- **Pegasus** stands 6 ft tall, has a horse's body, a horse's head, four legs, and two wings.
- **Centaur** stands 6 ft tall, has a horse's body with a human torso, a male human head, and four legs.
- **Griffin** stands 4–6 ft tall, has a lion's body, an eagle's head, four legs, two wings, fur on its body, and feathers on its head and wings.
- **Dragon** can grow to several hundred feet, has a snake-like body, from 1 to 3 reptile-like heads, four legs, scales, and breathes fire.
- **Chimera** stands 6 ft tall, has a goat's body, snake's tail, four legs, a lion's head, fur on its body and head, scales on its tail, and breathes fire.
- **Hydra** is several hundred feet long, has a long body with four legs and a spiked tail, 100 snake heads, scales, and is poisonous.

Analysis

1. **Identify** the characteristics that you think are the most useful for grouping the organisms into separate groups.

   Answers will vary.

2. **Classify** the organisms into at least three groups based on the characteristics you think are most important.

   Answers will vary. Students may group the creatures according to size, presence of wings, or other features.

3. **Evaluate** the use of the biological species concept to classify these mythical organisms.

   The biological species concept cannot be used to classify them without knowledge of their breeding compatibility with other groups.

Name ______________________ Class ______________ Date ______________

Data Lab

DATASHEET FOR IN-TEXT LAB

Making a Cladogram

Background

A cladogram is a model that represents a hypothesis about the order in which organisms evolved from a common ancestor. Scientists construct a cladogram by first analyzing characters in a data table. The absence of a vascular system and the absence of seeds is ancestral. Use the data to construct a cladogram on a separate sheet of paper.

Characters		
Plants	**Seeds**	**Vascular system**
Horsetails	No	Yes
Liverworts	No	No
Pine trees	Yes	Yes

Analysis

1. **Identify** the out-group.

   Liverworts are the out-group.

2. **Name** the least common derived trait.

   Seeds are the least common derived group.

3. **List** the order in which the plants in the table would be placed on a cladogram.

   liverwort, horsetail, pine tree

Name ______________________ Class ______________ Date ______________

Skills Practice Lab

DATASHEET FOR IN-TEXT LAB

Making a Dichotomous Key

SKILLS

- Identifying and comparing
- Organizing data

OBJECTIVES

- **Identify** objects using dichotomous keys.
- **Design** a dichotomous key for a group of objects.

MATERIALS

- 6 to 10 objects found in the classroom (e.g., shoes, books, writing instruments)
- stick-on labels
- pencil

Before You Begin

One way to identify an unknown organism is to use an **identification key,** which contains the major characteristics of groups of organisms. A **dichotomous key** is an identification key that contains pairs of contrasting descriptions. After each description, a key either directs the user to another pair of descriptions or identifies an object. In this lab, you will design and use a dichotomous key. A dichotomous key can be written for any group of objects.

1. Write a definition for each boldface term in the paragraph above.

____

____

____

____

2. Based on the objectives for this lab, write a question you would like to explore about making or using a dichotomous key.

Answers will vary. For example: How do you use a dichotomous key?

____

Procedure

PART A: USING A DICHOTOMOUS KEY

1. Use the **Key to Forest Trees** to identify the tree that produced each of the leaves shown. Identify one leaf at a time. Always start with the first pair of statements (**1***a* and **1***b*). Follow the direction beside the statement that describes the leaf. Proceed through the key until you get to the name of a tree.

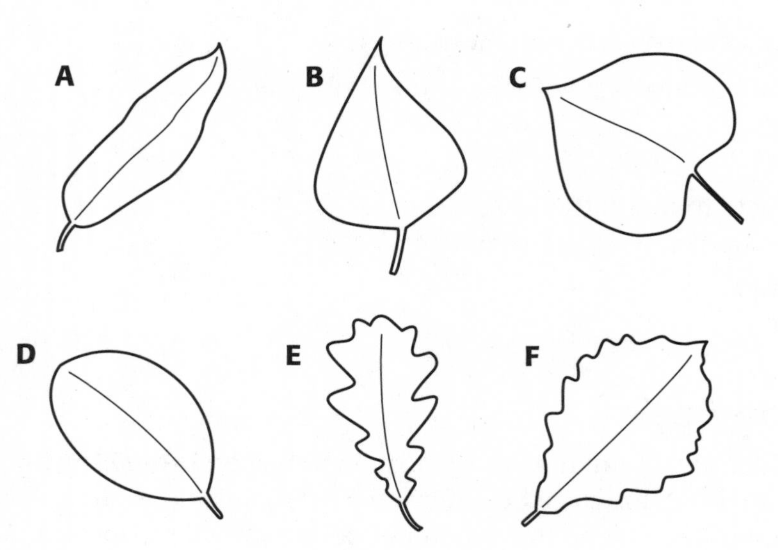

Key to Forest Trees		
1*a*	Leaf edge has no teeth, waves, or lobes	go to **2**
1*b*	Leaf edge has teeth, waves, or lobes	go to **3**
2*a*	Leaf has a bristle at its tip	**shingle oak**
2*b*	Leaf has no bristle at its tip	go to **4**
3*a*	Leaf edge is toothed	**Lombardy poplar**
3*b*	Leaf edge has waves or lobes	go to **5**
4*a*	Leaf is heart-shaped	**red bud**
4*b*	Leaf is not heart-shaped	**live oak**
5*a*	Leaf edge has lobes	**English oak**
5*b*	Leaf edge has waves	**chestnut oak**

PART B: DESIGN A DICHOTOMOUS KEY

2. Work with the members of your lab group to design a dichotomous key using the materials listed for this lab.

You Choose
As you design your key, decide the following:

- **a.** what question you will explore
- **b.** what objects your key will identify
- **c.** how you will label personal property
- **d.** what distinguishing characteristics the objects have
- **e.** which characteristics to use in your key
- **f.** how you will organize the data you will need for writing your key

3. Before you begin writing your key, have your teacher approve the objects your group has decided to work with.
4. Using the **Key to Forest Trees** as a guide, write a key for the objects your group selected. Remember, a dichotomous key includes pairs of contrasting descriptions.
5. Use your key to explore one of the questions written for step 2 of **Before You Begin.**

   Answers will vary.

6. After each group has completed step 5, exchange keys and the objects they identify with another group. Use the key you receive to identify the objects. If the key does not work, return it to the group so corrections can be made.

PART C: CLEANUP

7. Clean up your work area and all lab equipment. Return lab equipment to its proper place. Wash your hands thoroughly before you leave the lab and after you finish all work.

Analyze and Conclude

1. **Drawing Conclusions** What tree produced each of the leaves shown in this lab?

   A. shingle oak; B. Lombardy poplar; C. red bud; D. live oak; E. English oak; and F. chestnut oak

2. Forming Hypotheses What other characteristics might be used to identify leaves with a dichotomous key?

Other leaf characteristics might include whether the leaf is compound or simple, whether the leaf is needlelike, the arrangement of leaves on the branches, and the leaf's vein pattern.

3. Analyzing Methods How was the key your group designed dichotomous?

The key was dichotomous because the descriptive statements were written as pairs of contrasting statements.

4. Evaluating Results Were you able to use another group's key to identify the objects for which it was written? If not, describe the problems you encountered.

Answers will vary. Students might have discovered that some of the opposing statements were not contrasting, that some of the objects were not correctly described, or that the key did not include a sufficient number of descriptions to distinguish all the objects from one another.

5. Analyzing Methods Does a dichotomous key begin with general descriptions and then proceed to more specific descriptions or vice versa? Explain your answer, giving an example from your key.

Dichotomous keys proceed from general characteristics to specific characteristics. Examples from keys will vary but should reflect this gradation.

6. Further Inquiry Write a new question about making or using keys that could be explored with another investigation.

Answers will vary. For example: Can the members of any kind of group be identified using a dichotomous key?

Classification

Teacher Notes

TIME REQUIRED One 45-minute period

SKILLS ACQUIRED

Classifying
Communicating
Identifying and recognizing patterns
Organizing and analyzing data

RATINGS Easy ← 1 2 3 4 → Hard

Teacher Prep-1
Student Setup-1
Concept Level-3
Cleanup-1

THE SCIENTIFIC METHOD

Communicate the Results Students make a dichotomous key that reflects their classification scheme for imaginary species living in the future.

Analyze the Results Analysis question 3 requires students to analyze their results.

Draw Conclusions Conclusions questions 1 and 2 require students to draw conclusions.

TECHNIQUES TO DEMONSTRATE

You may wish to use a collection of buttons that vary in shape, size, and color to demonstrate how objects can be placed into successively smaller groups, based on the objects' similar characteristics.

TIPS AND TRICKS

Preparation

This lab may be done individually or in groups of two, as in-class work or outside the class.

Discuss with students the science of taxonomy and the use of classification keys.

You may want to review the terms *heterotrophic*, *autotrophic*, *chemosynthetic*, *anaerobic*, and *parasitic*.

Procedure

Point out to students that the species names they create must include the names of the genera in their classification scheme.

After students do step 5, ask them to identify the kingdom, phylum, and genus to which each species belongs.

Students should end this lab with the understanding that classification systems using scientific names improve the effectiveness of communication regarding biological diversity.

Exploration Lab

OBSERVATION

Classification

Taxonomy is the science of identifying and classifying living things. In taxonomy, organisms are grouped based on their similarities. By classifying organisms according to certain traits, biologists are able to determine which organisms may be related. As new technology arises, scientists can reevaluate the way organisms have been grouped. For example, DNA sequencing has given biologists an important characteristic with which to group organisms.

Biologists think there may be as many as 40 million different species of organisms on Earth. A *species* is a group of organisms capable of mating with one another to produce fertile offpsring. Only about 1.5 million of the species on Earth have been named and described. Even though each species is different from all others, there are many similar characteristics among species. These similarities have made it possible for biologists to classify species into larger groups.

The naming of species based on their classification is very important because it makes communication about organisms more efficient. Biologists use the *Linnaean binomial system* to name a species. Each species is identified by two Latin names, or a *binomial*. The first name of the binomial identifies the *genus*, or group of species more closely related to one another than to any other group. The second name identifies a particular species within the genus. For example, grizzly bears belong to the species *Ursus arctos*. Polar bears are called *Ursus maritimus*. Both belong to the genus *Ursus* but are separate species.

In this lab, you will classify and give species names to imaginary organisms according to their characteristics. You will also create a dichotomous key that can be used to identify the organisms.

OBJECTIVES

Classify imaginary organisms according to their characteristics.

Create species names for the organisms.

Develop a dichotomous key for identifying the organisms.

MATERIALS

• pen or pencil

Procedure

1. The year is 2500. All bodies of water on Earth have become extremely polluted. Because of the pollution, almost every species of animal, plant, fungus, protist, bacteria, and virus that existed in the early 2000s has become extinct. However, because of natural selection, genetic engineering, and selective breeding programs, particular areas of the planet have been repopulated. The organisms now present are very different variations of the organisms that once existed.

The ten major species of organisms include the following:

A. chemosynthetic aquatic organisms that live on water contaminants

B. heterotrophic, aquatic organisms who work on “aqua farms”

C. terrestrial organisms with green, photosynthetic hair

D. anaerobic, heterotrophic, space travelers who must wear deoxygenated suits to survive on Earth

E. photosynthetic, legless space travelers who visit Earth every month and who have armlike tentacles

F. green-skinned, photosynthetic, transparent-haired terrestrial organisms

G. parasitic winged organisms that feed off of any terrestrial life form

H. decomposers that feed on pollution and solid waste

I. heterotrophic, microscopic decomposers that feed on aquaculture crops or any photosynthetic organism

J. heterotrophic aquatic crop, grown on the aqua farm, that feeds on bacteria or anything autotrophic

2. You have just arrived from another planet to classify these ten species. Create a classification scheme for the species to help you and others study them. Use the taxonomic tiers *kingdom, phylum, and genus.* First, group the species into two kingdoms, based on the organisms’ characteristics. Record the names of your kingdoms in the two boxes on the next page. Also record the letters of the species (A–J) that belong to each kingdom.

3. On the next page draw the number of boxes that you will need for your scheme. Then record in the boxes the names of your phyla and the letters of the species that belong to each phylum. Use a vertical line to connect each phylum to the kingdom to which it belongs.

Kingdom

Phylum

Students' classification schemes will vary partly because the criteria they choose for placing the organisms into the two kingdoms. For example, kingdoms classified as aerobic or anaerobic will have different phyla than kingdoms classified as aquatic or terrestrial.

Genus

4. Repeat step 3 for the genera in your classification scheme, except connect each genus to the phylum to which it belongs.
5. Create a species name for each of the ten species, using the Linnaean binomial system. Be sure to capitalize the genus name only. Write these names in the table on the next page.

Name ______________________ Class ______________ Date ______________

Classification *continued*

TABLE 1 SPECIES LIVING IN THE YEAR 2500

Species description	Species name
A. chemosynthetic aquatic organisms	
B. heterotrophic aquatic farmer	
C. photosynthetic-haired terrestrial organisms	
D. anaerobic space travelers	
E. photosynthetic legless space travelers	
F. photosynthetic green-skinned terrestrial organisms	
G. parasitic winged organisms	
H. pollution and waste decomposers	
I. microscopic decomposers	
J. heterotrophic aquatic crop	

6. On the following page, create a dichotomous key for identifying the ten species. The key will need to be understood by your alien colleagues who will be identifying the species on their summer visits to Earth. To create a key, write two opposing descriptive statements that will divide the species into two groups. Start with a pair of statements that will separate the ten species into the two kingdoms you devised. At the end of each statement, write the number of the next set of opposing statements that will further describe and divide the species in each group. If the description of a species is complete, write the name of the species. Continue to write opposing statements until all the species are identified. For example:

Dichotomous Key	
1. Organism is autotrophic.	Go to 2.
Organism is not autotrophic.	Go to 3.
2. Organism contains chlorophyll.	Go to 6.
Organism does not contain chlorophyll.	*Name of species*

DICHOTOMOUS KEY FOR SPECIES LIVING IN THE YEAR 2500

7. Exchange your completed key with another student. Using the new key, identify that student's names for the ten species.

Analysis

1. Classifying What characteristic did you use to group the species into two kingdoms?

Answers will vary. Consideration of whether a species is autotrophic or heterotrophic is one characteristic that students may use. Other criteria are aerobic/anaerobic, aquatic/terrestrial (or extraterrestrial), and number of appendages.

2. Describing Events Were you able to follow the dichotomous key created by the other student? What problems, if any, did you have with the key you tested?

If the student could not follow the key, he or she should specify the problems encountered.

3. Analyzing Data Compare your key with the keys developed by other students. How are they similar? How are they different?

Answers will vary.

Conclusions

1. Drawing Conclusions Can more than one species occupy the same final place on your key? Why?

No, the point of a dichotomous key is to use enough categories to classify organisms as distinct species.

2. Evaluating Methods What advantage does the dichotomous key you developed have over descriptions of the ten species?

The key is less subjective than descriptions and allows for more accurate grouping of organisms. Thus, it can be used to identify the species names of unfamiliar organisms.

Extensions

1. **Building Models** Link the ten species listed in **Table 1** in two possible food chains, beginning with a producer (autotroph) and ending with a decomposer. Each food chain should have a minimum of four species. The same single species may appear in both food chains.
2. **Research and Communications** With a small group of your classmates, discuss the environmental issue of how pollution affects species and their evolution. How might genetic engineering or selective breeding programs change the course of the evolution of a species?

Analyzing Amino-Acid Sequences

Teacher Notes

TIME REQUIRED One 45-minute period

SKILLS ACQUIRED

Collecting data
Inferring
Interpreting
Organizing and analyzing data

RATINGS Easy ← 1 2 3 4 → Hard

Teacher Prep–2
Student Setup–2
Concept Level–3
Cleanup–1

THE SCIENTIFIC METHOD

Make Observations Students observe differences among the amino-acid sequences of various organisms.

Analyze the Results Analysis questions 1 and 2 require students to analyze their results.

Draw Conclusions Conclusions questions 1–3 require students to draw conclusions from their data.

SAFETY CAUTIONS

Discuss all safety symbols with students.

TIPS AND TRICKS

This lab works best in groups of two students.

For each group, make a photocopy of Figure 1 and Figure 2.

The lab shows only differences in two macromoles found in humans and other vertebrates, it does not show evolutionary relationships. Thus, for Extensions item 1, the cladogram that students make from their data should show branches extending from a single line that leads to humans. The order of the branches should reflect the order of the vertebrates listed in Table 1.

You may wish to have students also compare the nonhuman vertebrate amino-acid sequences to each other.

Name ______________________ Class ______________ Date ______________

Skills Practice Lab

OBSERVATION

Analyzing Amino-Acid Sequences

The biochemical comparison of proteins is a technique used to determine evolutionary relationships among organisms. Proteins consist of chains of amino acids. The sequence, or order, of the amino acids in a protein determines the type and nature of the protein. In turn, the sequence of amino acids in a protein is determined by the sequence of nucleotides in a gene. A change in the DNA nucleotide sequence (mutation) of a gene that codes for a protein may result in a change in the amino-acid sequence of the protein.

Biochemical evidence of evolution compares favorably with structural evidence of evolution. Even organisms that appear to have few physical similarities may have similar sequences of amino acids in their proteins and be closely related through evolution. Many researchers believe that the greater the similarity in the amino-acid sequences of two organisms, the more closely related they are in an evolutionary sense. Conversely, the greater the time that organisms have been diverging from a common ancestor, the greater the differences that can be expected in the amino-acid sequences of their proteins.

Two proteins are commonly studied in attempting to deduce evolutionary relationships from differences in amino-acid sequences. One is cytochrome *c*, and the other is hemoglobin. *Cytochrome* c is a protein used in cellular respiration and found in the mitochondria of many organisms. *Hemoglobin* is the oxygen-carrying molecule found in red blood cells.

In this lab, you will compare portions of human cytochrome c and hemoglobin molecules with the same portions of those molecules in other vertebrates. You will determine the differences in the amino-acid sequences of the molecules and deduce the evolutionary relationships among the vertebrates.

OBJECTIVES

Identify differences in the amino-acid sequences of the cytochrome *c* and hemoglobin molecules of several vertebrates.

Infer the evolutionary relationships among several vertebrates by comparing the amino-acid sequences of the same protein in those vertebrates.

MATERIALS

- scissors
- photocopy of cytochrome *c* amino-acid sequences
- photocopy of hemoglobin amino-acid sequences

Procedure

PART 1: CYTOCHROME *C*

A cytochrome c molecule consists of a chain of 104 amino acids. **Figure 1** shows the amino-acid sequences in a section of the cytochrome *c* molecules of eight vertebrates. The numbers refer to the positions of these amino acids.

1. Using a photocopy of the chart in **Figure 1,** cut out each vertebrate's amino-acid sequence. You should have eight amino-acid sequence strips.

FIGURE 1 CYTOCHROME *C* AMINO-ACID SEQUENCES

AA#	Horse	Chicken	Frog	Human	Shark	Turtle	Monkey	Rabbit
42	GLN	GLN	GLN	GLN	GLN	GLN	GLN	GLN
43	ALA	ALA	ALA	ALA	ALA	ALA	ALA	ALA
44	PRO	GLU	ALA	PRO	GLN	GLU	PRO	VAL
46	PHE	PHE	PHE	TYR	PHE	PHE	TYR	PHE
47	THR	SER	SER	SER	SER	SER	SER	SER
49	THR	THR	THR	THR	THR	THR	THR	THR
50	ASP	ASP	ASP	ALA	ASP	GLU	ALA	ASP
53	LYS	LYS	LYS	LYS	LYS	LYS	LYS	LYS
54	ASN	ASN	ASN	ASN	ASN	ASN	ASN	ASN
55	LYS	LYS	LYS	LYS	LYS	LYS	LYS	LYS
56	GLY	GLY	GLY	GLY	GLY	GLY	GLY	GLY
57	ILE	ILE	ILE	ILE	ILE	ILE	ILE	ILE
58	THR	THR	THR	ILE	THR	THR	ILE	THR
60	LYS	GLY	GLY	GLY	GLN	GLY	GLY	GLY
61	GLU	GLU	GLU	GLU	GLN	GLU	GLU	GLU
62	GLU	ASP	ASP	ASP	GLU	GLU	ASP	ASP
63	THR	THR	THR	THR	THR	THR	THR	THR
64	LEU	LEU	LEU	LEU	LEU	LEU	LEU	LEU
65	MET	MET	MET	MET	ARG	MET	MET	MET
66	GLU	GLU	GLU	GLU	ILE	GLU	GLU	GLU
100	LYS	ASP	SER	LYS	LYS	ASP	LYS	LYS
101	ALA	ALA	ALA	ALA	THR	ALA	ALA	ALA
102	THR	THR	CYS	THR	ALA	THR	THR	THR
103	ASN	SER	SER	ASN	ALA	SER	ASN	ASN
104	GLU	LYS	LYS	GLU	SER	LYS	GLU	GLU

2. Compare the amino-acid sequence of human cytochrome *c* with that of each of the other seven vertebrates by aligning the appropriate strips side by side.
3. For each nonhuman vertebrate's sequence, count the number of amino acids that differ from those in the human sequence. Write the number of differences next to that vertebrate's name on the strip.
4. When you have completed your comparisons, transfer your data to **Table 1.** As you do, list the seven nonhuman vertebrates in order from fewest differences to most differences.

TABLE 1 CYTOCHROME C AMINO-ACID SEQUENCE DIFFERENCES

Vertebrate	Number of differences from human cytochrome c
monkey	0
rabbit	4
horse	6
chicken	7
turtle	8
frog	8
shark	13

PART 2: HEMOGLOBIN

5. **Figure 2** shows the amino-acid sequences in corresponding parts of the hemoglobin molecules of five vertebrates. The parts of the chains shown are from amino acid number 87 to amino acid number 116, within a total sequence of 146 amino acids. Using a photocopy of the chart in **Figure 1,** cut out each vertebrate's amino-acid sequence to form five amino-acid sequence strips.

Name ______________________ Class ______________ Date ______________

Analyzing Amino-Acid Sequences *continued*

FIGURE 2 HEMOGLOBIN PROTEIN AMINO-ACID SEQUENCES

AA#	Human	Chimpanzee	Gorilla	Monkey	Horse
87	THR	THR	THR	GLN	ALA
88	LEU	LEU	LEU	LEU	LEU
89	SER	SER	SER	SER	SER
90	GLU	GLU	GLU	GLU	GLU
91	LEU	LEU	LEU	LEU	LEU
92	HIS	HIS	HIS	HIS	HIS
93	CYS	CYS	CYS	CYS	CYS
94	ASP	ASP	ASP	ASP	ASP
95	LYS	LYS	LYS	LYS	LYS
96	LEU	LEU	LEU	LEU	LEU
97	HIS	HIS	HIS	HIS	HIS
98	VAL	VAL	VAL	VAL	VAL
99	ASP	ASP	ASP	ASP	ASP
100	PRO	PRO	PRO	PRO	PRO
101	GLU	GLU	GLU	GLU	GLU
102	ASN	ASN	ASN	ASN	ASN
103	PHE	PHE	PHE	PHE	PHE
104	ARG	ARG	LYS	LYS	ARG
105	LEU	LEU	LEU	LEU	LEU
106	LEU	LEU	LEU	LEU	LEU
107	GLY	GLY	GLY	GLY	GLY
108	ASN	ASN	ASN	ASN	ASN
109	VAL	VAL	VAL	VAL	VAL
110	LEU	LEU	LEU	LEU	LEU
111	VAL	VAL	VAL	VAL	ALA
112	CYS	CYS	CYS	CYS	LEU
113	VAL	VAL	VAL	VAL	VAL
114	LEU	LEU	LEU	LEU	VAL
115	ALA	ALA	ALA	ALA	ALA
116	HIS	HIS	HIS	HIS	ARG

6. Use the strips you cut out to compare the amino-acid sequence of human hemoglobin with that of each of the other four vertebrates.
7. For each nonhuman vertebrate's sequence, count the number of amino acids that differ from the human sequence.
8. Write the number of differences next to that vertebrate's name on the strip.
9. When you have completed your comparisons, transfer your data to **Table 2.** Be sure to list the four vertebrates in order from fewest differences to most differences.

TABLE 2 HEMOGLOBIN AMINO-ACID SEQUENCE DIFFERENCES

Vertebrate	Number of differences from human hemoglobin
chimpanzee	**0**
gorilla	**1**
monkey	**2**
horse	**5**

Analysis

1. **Identifying Relationships** According to the data in **Table 1,** which vertebrate is most closely related to humans? Which is least closely related to humans?

   The monkey is most closely related to humans. The shark is least closely related.

2. **Identifying Relationships** According to the data in **Table 2,** which vertebrate is most closely related to humans? Least closely related?

   The chimpanzee is most closely related; the horse is least closely related.

3. **Identifying Relationships** If the amino-acid sequences in gorillas and humans are similar, are the nucleotide sequences of their DNA also similar? Why?

   Yes, their nucleotide sequences are similar because the amino-acid sequences of proteins are encoded by the nucleotide sequences of DNA.

Conclusions

1. **Evaluating Methods** Can you deduce from the data in **Table 1** that the chicken and the horse are closely related to each other? Why or why not?

   No, the data in the table resulted from comparing chickens and horses with humans, not with each other.

2. **Drawing Conclusions** According to the data listed in **Table 2,** what conclusion can you make about how closely the three primates—chimpanzee, gorilla, and monkey—are related to each other?

   The chimpanzee is more closely related to the gorilla than to the monkey.

3. **Applying Conclusions** In what way do proteins behave like molecular clocks?

   Proteins behave like molecular clocks in that they change gradually over time due to mutations. The number of changes in their amino-acid sequences might be considered a measure of the passage of time. The greater the number of changes, the more time has passed.

Extensions

1. **Building Models** Use your data in **Table 1** to make a cladogram that shows the evolutionary relationships between humans and the seven vertebrates listed in the table.
2. **Research and Communications** Research how biologists determine the amino-acid sequence of a protein molecule.

Answer Key

Directed Reading

SECTION: CATEGORIES OF BIOLOGICAL CLASSIFICATION

1. taxonomy
2. structural
3. binomial nomenclature
4. species
5. genus
6. *Quercus phellos*
7. red oak
8. genus
9. genus
10. *Erithacus rubicula*
11. *H. sapiens*
12. 3
13. 4
14. 8
15. 6
16. 5
17. 2
18. 7
19. 1
20. Biologists named organisms by adding descriptive phrases to the name of the genus. These phrases sometimes consisted of 12 or more Latin words and were called polynomials.
21. The first letter of the genus name is always capitalized, and the first letter of the second word is always lowercase. Scientific names are italicized or underlined. After the first use of the full scientific name, the genus name can be abbreviated as a single letter.

SECTION: HOW BIOLOGISTS CLASSIFY ORGANISMS

1. a
2. c
3. d
4. b
5. b
6. d
7. Reproductive isolation occurs when a barrier separates two or more groups of organisms and prevents them from interbreeding.
8. When individuals of different species interbreed and produce offspring, the offspring are called hybrids.
9. In convergent evolution, organisms evolve similar features independently. These features are called analogous characters.
10. Cladistics is a system of taxonomy that reconstructs phylogenies by focusing on derived traits, which are unique characteristics found in a particular group of organisms.
11. the evolutionary history of a species
12. in order to accurately reconstruct the evolutionary history of an organism
13. A cladogram is a branching diagram that shows the evolutionary relationships among groups of organisms.
14. d
15. c
16. b
17. d
18. a

Active Reading

SECTION: CATEGORIES OF BIOLOGICAL CLASSIFICATION

1. form and structure
2. The groups become more and more inclusive, encompassing more members who share a fewer number of traits.
3. six; Archaebacteria, Eubacteria, Protista, Fungi, Plantae, and Animalia
4. **a.** Phyla
   b. A kingdom contains phyla.
   c. Classes
   d. A phylum contains classes.
   e. Orders
   f. A class contains orders.
   g. Families
   h. An order contains families.
   i. Genera.
   j. A family contains genera.
   k. A genus contains species.
5. d

SECTION: HOW BIOLOGISTS CLASSIFY ORGANISMS

1. by inferring relationships based on similarities derived from a common ancestor
2. the sequence in which different groups of organisms evolved
3. Cladistics focuses on derived characters, which are a set of unique characteristics found in a particular group of organisms.
4. c

Vocabulary Review

1. phylum
2. evolutionary systematics
3. cladistics
4. phylogeny
5. class
6. order
7. family
8. analogous character
9. convergent evolution
10. cladogram
11. domain
12. genus
13. taxonomy
14. biological species
15. derived characters
16. binomial nomenclature
17. kingdom
18. phylogenetic tree

Science Skills

ORGANIZING INFORMATION

1. **a.** bat, chicken, eagle, frog, grasshopper, horse, polar bear, rabbit, spider
   b. frog, goldfish, octopus, whale
   c. bat, eagle, frog, goldfish, octopus, polar bear, spider, whale, chicken
   d. grasshopper, horse, rabbit
   e. chicken, eagle, frog, goldfish, grasshopper, octopus, spider
   f. bat, horse, polar bear, rabbit, whale
   g. chicken, eagle, frog, goldfish
   h. chicken, eagle
   i. bat, horse, polar bear, rabbit
   j. whale
2. Group 4; More characteristics and combinations of characteristics are considered in Group 4 than in any of the other groups. Group 4 allows the organisms to be categorized in more specific terms. As a result, Group 4 is the only classification system that separates birds, land mammals, and aquatic mammals. Although the goldfish and frog belong to two different major taxonomic groups, their grouping is more appropriate than a group that also includes a polar bear and a spider, as in Group 2.
3. Yes, frogs live on land and in the water at different stages of their life cycle. The octopus, the spider, and the grasshopper lay eggs, but they are invertebrates.

Concept Mapping

1. biological classification
2. phylogenetics
3. domain
4. kingdom
5. phylum
6. class
7. order
8. genus
9. species
10. binomial nomenclature

Critical Thinking

1. b
2. d
3. a
4. c
5. c
6. e
7. a
8. d
9. b
10. c
11. b
12. a
13. a, j
14. k, f
15. l, h
16. g, b
17. e, d
18. c, i
19. b
20. b

Test Prep Pretest

1. b
2. d
3. a
4. d
5. c
6. b
7. b
8. c
9. d
10. d
11. a
12. d
13. binomial nomenclature
14. species
15. biological species

16. reproductive barriers
17. Animalia
18. convergent evolution
19. homologous
20. phylogeny
21. Homologous characters are present in different taxonomic groups but have arisen from a recent common ancestor. The wing of a bat, a person's arm, and a bird's wing are homologous structures that contain the same basic arrangement of bones as a common vertebrate ancestor. Analogous characters have evolved independently through convergent evolution. Both a bird's wing and an insect's wing function in flight but have completely different physical structures.
22. Cladistics would most likely be the first classification system used. Though the scientist could identify the physical characteristics of the plants, the scientist would probably not yet understand the role those traits play in the plants' evolutionary history.
23. Mayr's concept of biological species as organisms that naturally breed only with other members of their own species begins to break down when considering closely related species with incomplete reproductive barriers and organisms that reproduce asexually.
24. 3 and 4
25. D

Quiz

SECTION: CATEGORIES OF BIOLOGICAL CLASSIFICATION

1. b
2. c
3. d
4. c
5. a
6. c
7. d
8. b
9. e
10. a

SECTION: HOW BIOLOGISTS CLASSIFY ORGANISMS

1. d
2. a
3. b
4. b
5. c
6. c
7. e
8. a
9. b
10. d

Chapter Test (General)

1. c
2. b
3. d
4. a
5. b
6. d
7. a
8. c
9. b
10. d
11. c
12. a
13. d
14. c
15. d
16. e
17. a
18. b
19. f
20. c

Chapter Test (Advanced)

1. c
2. b
3. d
4. c
5. b
6. a
7. d
8. b
9. c
10. d
11. b
12. d
13. d
14. b
15. c
16. f
17. g
18. e
19. b
20. a
21. c
22. d
23. The scientific name of an organism gives biologists a common way of communicating regardless of their native language. One species may have many common names, and many species may have the same common name. Each species has only one scientific name in the binomial nomenclature system.
24. The birds are not the same species because their reproductive cycles are not the same. The varying times of year when they breed in nature is a reproductive barrier. They are reproductively isolated.
25. In convergent evolution, similar structures called analogous characters, such as the wings of birds and insects, evolve in unrelated species. The species evolved through selective pressures in similar environments where the character was an advantage. Analogous characters have the same function, such as flight, but evolve independently.

Lesson Plan

Section: Categories of Biological Classification

Pacing

Regular Schedule: **with lab(s):** 4 days **without lab(s):** 3 days

Block Schedule: **with lab(s):** 2 days **without lab(s):** 1 1/2 days

Objectives

1. Describe Linnaeus's role in developing the modern system of naming organisms.
2. Summarize the scientific system for naming a species.
3. List the seven levels of biological classification.

National Science Education Standards Covered

UNIFYING CONCEPTS AND PROCESSES

UCP1: Systems, order, and organization

UCP2: Evidence, models, and explanation

UCP5: Form and function

SCIENCE AS INQUIRY

SI1: Abilities necessary to do scientific inquiry

SI2: Understandings about scientific inquiry

HISTORY AND NATURE OF SCIENCE

HNS1: Science as a human endeavor

HNS2: Nature of scientific knowledge

HNS3: Historical perspectives

LIFE SCIENCE: BIOLOGICAL EVOLUTION

LSEvol1: Species evolve over time.

LSEvol2: The great diversity of organisms is the result of more than 3.5 billion years of evolution.

LSEvol3: Natural selection and its evolutionary consequences provide a scientific explanation for the fossil record of ancient life forms, as well as for the striking molecular similarities observed among the diverse species of living organisms

LSEvol4: The millions of different species of plants, animals, and microorganisms that live on earth today are related by descent from common ancestors.

LSEvol5: Biological classifications are based on how organisms are related.

KEY	
SE = Student Edition	**TE** = Teacher Edition
CRF = Chapter Resource File	

Block 1

CHAPTER OPENER *(45 minutes)*

_ **Quick Review,** SE. Students answer questions covered in previous sections of the textbook as preparation for the chapter content. **(GENERAL)**

_ **Reading Activity,** SE. Students study the figures and captions in each section. For each figure, students write a question that can be answered by referring to the figure and its caption. **(GENERAL)**

_ **Using the Figure,** TE. Students answer questions about the chapter opener photograph. **(GENERAL)**

_ **Opening Activity**, TE. Students study two plants with obvious differences. Then students come up with one word that describes both plants and one word that distinguishes the two plants. **(BASIC)**

Block 2

FOCUS *(5 minutes)*

_ **Bellringer Transparency.** Use this transparency as students enter the classroom and find their seats. **(GENERAL)**

MOTIVATE *(10 minutes)*

_ **Discussion/Question**, TE. Students discuss the problem of two people having the same name and ways to differentiate the two people. **(BASIC)**

TEACH *(30 minutes)*

_ **Teaching Transparency, Section Outline.** Use this transparency to give students a framework for the information in this section. **(GENERAL)**

_ **Reading Skill Builder**, Brainstorming, TE. Students brainstorm things that are categorized using a hierarchical classification system. **(GENERAL)**

_ **Demonstration**, TE. Students list common names for *Felis concolor* and then discuss other examples of misleading common names. **(GENERAL)**

_ **Teaching Transparency, Biological Hierarchy of Classification.** Use this transparency to introduce the eight levels of classification used in the modern scientific classification of organisms. Help students understand what the white and blue-green ovals represent. **(GENERAL)**

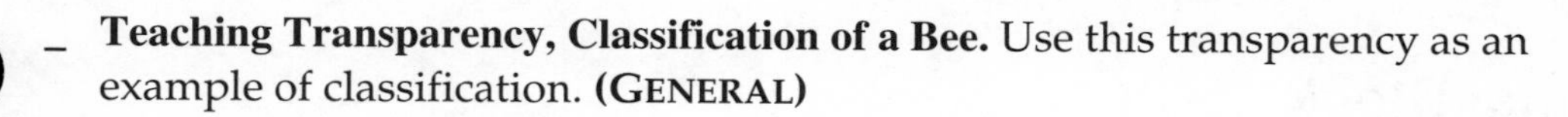

- **Teaching Transparency, Classification of a Bee.** Use this transparency as an example of classification. **(GENERAL)**

HOMEWORK

- **Active Reading Worksheet, Categories of Biological Classification, CRF.** Students read a passage related to the section topic and answer questions. **(GENERAL)**
- **Directed Reading Worksheet, Categories of Biological Classification, CRF.** Students complete the exercises in this worksheet to help them understand the material as they read the section. **(BASIC)**
- **Basic Skills Worksheet, Words and Word Roots, One-Stop Planner.** Students learn about words roots in this worksheet. **(GENERAL)**

Block 3

TEACH *(30 minutes)*

- **Demonstration**, TE. Students compare photos and scientific names of different organisms. Then they suggest why each organism was given its particular scientific name. **(GENERAL)**
- **Activity**, Comparing Classifications, TE. Students construct a graphic organizer to compare humans, chimpanzees, and gorillas in a classification hierarchy beginning with the kingdom Animalia. **(GENERAL)**
- **Quick Lab,** Using Field Guides, SE. Students to come. **(GENERAL)**
- **Datasheets for In-Text Labs, Using Field Guides, CRF.**

CLOSE *(15 minutes)*

- **Reteaching,** TE. Students research the meanings of given scientific names and the common names of the organisms represented by each scientific name. **(BASIC)**

HOMEWORK

- **Quiz,** TE. Students answer questions that review the section material. **(GENERAL)**
- **Section Review,** SE. Assign questions 1–4 for review, homework, or quiz. **(GENERAL)**
- **Alternative Assessment**, TE. Students determine what is wrong with each entry on a list of scientific names. **(GENERAL)**
- **Quiz,** CRF. This quiz consists of ten multiple choice and matching questions that review the section's main concepts. **(BASIC) Also in Spanish.**

Optional Block

LAB *(45 minutes)*

_ **Exploration Lab, Classification, CRF.** Students classify and give species names to imaginary organisms according to their characteristics. They also create a dichotomous key that can be used to identify the organisms. **(GENERAL)**

Other Resource Options

_ **Internet Connect.** Students can research Internet sources about Naming Species with SciLinks Code HX4127.

_ **Internet Connect.** Students can research Internet sources about Systems of Classification with SciLinks Code HX4173.

_ **Internet Connect.** Students can research Internet sources about Taxonomy with SciLinks Code HX4174.

_ **Internet Connect.** Students can research Internet sources about Classification with SciLinks Code HX4044.

_ **go.hrw.com.** For worksheets, videos, and other teaching aids related to this chapter, visit the HRW Web site and type in the keyword HX4 CLS.

_ **CNN Science in the News, Video Segment 11 Something Worth Saving.** This video segment is accompanied by a **Critical Thinking Worksheet**.

_ **CNN Student News.** Find the latest news, lesson plans, and activities related to important scientific events at **cnnstudentnews.com**.

Lesson Plan

Section: How Biologists Classify Organisms

Pacing

Regular Schedule:	**with lab(s):** 3 days	**without lab(s):** 2 days
Block Schedule:	**with lab(s):** 1 1/2 days	**without lab(s):** 1 day

Objectives

1. List the characteristics that biologists use to classify organisms.
2. Summarize the biological species concept.
3. Relate analogous structures to convergent evolution.
4. Describe how biologists use cladograms to determine evolutionary histories.

National Science Education Standards Covered

UNIFYING CONCEPTS AND PROCESSES

UCP1: Systems, order, and organization

UCP4: Evolution and equilibrium

UCP5: Form and function

SCIENCE AS INQUIRY

SI1: Abilities necessary to do scientific inquiry

SI2: Understandings about scientific inquiry

HISTORY AND NATURE OF SCIENCE

HNS1: Science as a human endeavor

HNS2: Nature of scientific knowledge

HNS3: Historical perspectives

LIFE SCIENCE: THE MOLECULAR BASIS OF HEREDITY

LSGene1: In all organisms, the instructions for specifying the characteristics of the organisms are carried in DNA.

LSGene2: Most of the cells in a human contain two copies of each of the 22 different chromosomes. In addition there is a pair of chromosomes that determine sex.

LSGene3: Changes in DNA (mutations) occur spontaneously at low rates.

LIFE SCIENCE: BIOLOGICAL EVOLUTION

LSEvol1: Species evolve over time.

LSEvol2: The great diversity of organisms is the result of more than 3.5 billion years of evolution.

LSEvol4: The millions of different species of plants, animals, and microorganisms that live on earth today are related by descent from common ancestors.

LSEvol5: Biological classifications are based on how organisms are related.

KEY
SE = Student Edition
TE = Teacher Edition
CRF = Chapter Resource File

Block 4

FOCUS *(5 minutes)*

_ **Bellringer Transparency.** Use this transparency as students enter the classroom and find their seats. **(GENERAL)**

MOTIVATE *(10 minutes)*

_ **Activity**, TE. Students work with a partner to group a handful of hardware into genera and species. **(GENERAL)**

TEACH *(30 minutes)*

_ **Teaching Transparency, Section Outline.** Use this transparency to give students a framework for the information in this section. **(GENERAL)**

_ **Data Lab,** Analyzing Taxonomy of Mythical Figures, SE. Students use descriptions of mythical figures to classify them into three groups. **(GENERAL)**

_ **Datasheets for In-Text Labs, Analyzing Taxonomy of Mythical Figures, CRF.**

_ **Using the Figure,** Figure 6, TE. Students discuss how cactuses and spurges can look so similar and yet not be closely related. **(GENERAL)**

_ **Reading Skill Builder,** Brainstorming, TE. Students brainstorm a list of analogous structures that are found in organisms. **(GENERAL)**

HOMEWORK

_ **Directed Reading Worksheet, How Biologists Classify Organisms, CRF.** Students complete the exercises in this worksheet to help them understand the material as they read the section. **(BASIC)**

_ **Active Reading Worksheet, How Biologists Classify Organisms, CRF.** Students read a passage related to the section topic and answer questions. **(GENERAL)**

Block 5

TEACH *(30 minutes)*

_ **Exploring Further,** Cladograms, SE. Students read this short article and dicuss the use of DNA analysis and traits in classifying organisms. **(GENERAL)**

_ **Data Lab,** Making a Cladogram, SE. Students use given data to construct a cladogram. **(GENERAL)**

_ **Datasheets for In-Text Labs, Making a Cladogram, CRF.**

_ **Teaching Transparency, Evolutionary Systematics and Cladistic Taxonomy.** Use this transparency to compare these two ways of classifying organisms. **(GENERAL)**

CLOSE *(15 minutes)*

_ **Reteaching,** TE. Students work with a partner to devise a table and a cladogram for five different animals. **(BASIC)**

_ **Quiz,** TE. Students answer questions that review the section material. **(GENERAL)**

HOMEWORK

_ **Alternative Assessment,** TE. Students prepare a graphic organizer to shoe the relationships of different geometric shapes. **(GENERAL)**

_ **Section Review,** SE. Assign questions 1–5 for review, homework, or quiz. **(GENERAL)**

_ **Science Skills Worksheet, CRF.** Students organize animals into four groups of classification. **(GENERAL)**

_ **Quiz, CRF.** This quiz consists of ten multiple choice and matching questions that review the section's main concepts. **(BASIC) Also in Spanish.**

_ **Modified Worksheet, One-Stop Planner.** This worksheet has been specially modified to reach struggling students. **(BASIC)**

_ **Critical Thinking Worksheet, CRF.** Students answer analogy-based questions that review the section's main concepts and vocabulary. **(ADVANCED)**

Optional Block

LAB *(45 minutes)*

_ **Skills Practice Lab,** Making a Dichotomous Key, SE. Students design and use a dichotomous key. **(GENERAL)**

_ **Datasheets for In-Text Labs, Making a Dichotomous Key, CRF.**

Other Resource Options

- **Skills Practice Lab, Analyzing Amino-Acid Sequences, CRF.** Students determine the differences in the amino-acid sequences of the molecules and deduce the evolutionary relationships among the vertebrates. **(GENERAL)**
- **Internet Connect.** Students can research Internet sources about Evolutionary Systematics with SciLinks Code HX4077.
- **Internet Connect.** Students can research Internet sources about Classification with SciLinks Code HX4044.
- **go.hrw.com.** For worksheets, videos, and other teaching aids related to this chapter, visit the HRW Web site and type in the keyword HX4 CLS.
- **CNN Science in the News, Video Segment 11 Something Worth Saving.** This video segment is accompanied by a **Critical Thinking Worksheet**.
- **CNN Student News.** Find the latest news, lesson plans, and activities related to important scientific events at **cnnstudentnews.com**.

Lesson Plan

End-of-Chapter Review and Assessment

Pacing

Regular Schedule: 2 days

Block Schedule: 1 day

KEY	
SE = Student Edition	**TE** = Teacher Edition
CRF = Chapter Resource File	

Block 6

REVIEW *(45 minutes)*

- **Study Zone,** SE. Use the Study Zone to review the Key Concepts and Key Terms of the chapter and prepare students for the Performance Zone questions. **(GENERAL)**
- **Performance Zone,** SE. Assign questions to review the material for this chapter. Use the assignment guide to customize review for sections covered. **(GENERAL)**
- **Teaching Transparency, Concept Mapping.** Use this transparency to review the concept map for this chapter. **(GENERAL)**

Block 7

ASSESSMENT *(45 minutes)*

- **Chapter Test, Classification of Organisms, CRF.** This test contains 20 multiple choice and matching questions keyed to the chapter's objectives. **(GENERAL) Also in Spanish.**
- **Chapter Test, Classification of Organisms, CRF.** This test contains 25 questions of various formats, each keyed to the chapter's objectives. **(ADVANCED)**
- **Modified Chapter Test, One-Stop Planner.** This test has been specially modified to reach struggling students. **(BASIC)**

Other Resource Options

- **Vocabulary Review Worksheet, CRF.** Use this worksheet to review the chapter vocabulary. **(GENERAL) Also in Spanish.**
- **Test Prep Pretest, CRF.** Use this pretest to review the main content of the chapter. Each question is keyed to a section objective. **(GENERAL) Also in Spanish.**
- **Test Item Listing for ExamView® Test Generator, CRF.** Use the Test Item Listing to identify questions to use in a customized homework, quiz, or test.
- **ExamView® Test Generator, One-Stop Planner.** Create a customized homework, quiz, or test using the HRW Test Generator program.

TEST ITEM LISTING

Classification of Organisms

TRUE/FALSE

1. ____ The Greeks developed the simple system used today for naming organisms.
   Answer: False Difficulty: I Section: 1 Objective: 1
2. ____ Carolus Linnaeus simplified the system for naming groups of organisms.
   Answer: True Difficulty: I Section: 1 Objective: 1
3. ____ Genus is the basic biological unit in the Linnaean system of classification.
   Answer: False Difficulty: I Section: 1 Objective: 2
4. ____ Species is a taxonomic category containing several genera.
   Answer: False Difficulty: I Section: 1 Objective: 2
5. ____ Two different organisms can have the same scientific name.
   Answer: False Difficulty: I Section: 1 Objective: 2
6. ____ Under the Linnaean system of classification, organisms are grouped based on form and structure.
   Answer: True Difficulty: I Section: 1 Objective: 2
7. ____ Linnaeus devised the eight levels into to which different groups of organisms can be classified.
   Answer: False Difficulty: I Section: 1 Objective: 3
8. ____ The least inclusive group to which an organism can be assigned is its kingdom.
   Answer: False Difficulty: I Section: 1 Objective: 3
9. ____ A species is the smallest taxonomic group into which an organism can be assigned.
   Answer: True Difficulty: I Section: 1 Objective: 3
10. ____ Each level of classification contains all organisms that share the same characteristics.
    Answer: True Difficulty: I Section: 2 Objective: 1
11. ____ All organisms in the kingdom Animalia are multicellular heterotrophs whose cells lack cell walls.
    Answer: True Difficulty: I Section: 2 Objective: 1
12. ____ Dogs and wolves cannot interbreed to produce fertile offspring.
    Answer: False Difficulty: I Section: 2 Objective: 2
13. ____ Interbreeding individuals of different species produce a hybrid.
    Answer: True Difficulty: I Section: 2 Objective: 2
14. ____ Classification provides strong evidence supporting Darwinian evolution.
    Answer: True Difficulty: I Section: 2 Objective: 3
15. ____ All traits are inherited from a common ancestor.
    Answer: False Difficulty: I Section: 2 Objective: 3
16. ____ Bat wings and bird wings are examples of analogous structures.
    Answer: False Difficulty: I Section: 2 Objective: 3
17. ____ Similar traits that evolve independently are the result of convergent evolution.
    Answer: True Difficulty: I Section: 2 Objective: 3
18. ____ Cladistics is used to determine the sequence in which different groups of organisms evolved.
    Answer: True Difficulty: I Section: 2 Objective: 4

19. ____ Cladograms are models that show the evolutionary relationship among homologous traits.

Answer: False Difficulty: I Section: 2 Objective: 4

20. ____ On a cladogram, all organisms share all traits.

Answer: False Difficulty: I Section: 2 Objective: 4

21. ____ An out-group is an unrelated organism on a cladogram.

Answer: False Difficulty: I Section: 2 Objective: 4

MULTIPLE CHOICE

22. Linnaeus's two-word system for naming organisms is called
    a. taxonomic evolution.
    b. Genus species.
    c. Greek polynomials.
    d. binomial nomenclature.

Answer: D Difficulty: I Section: 1 Objective: 1

23. Taxonomy is
    a. the study of life.
    b. the science of naming and classifying organisms.
    c. the evolutionary history of a species.
    d. the sequence in which different groups evolved.

Answer: B Difficulty: I Section: 1 Objective: 1

24. All scientific names must have
    a. two Latin words.
    b. the same species name.
    c. different genus names for organisms within the group.
    d. the same common name.

Answer: A Difficulty: I Section: 1 Objective: 2

25. The basic biological unit in the Linnaean system of classification is the
    a. kingdom.
    b. family.
    c. genus.
    d. species.

Answer: D Difficulty: I Section: 1 Objective: 2

26. An advantage of our scientific naming system is that
    a. common names mean the same in all countries.
    b. Latin names are easy to pronounce.
    c. biologists can communicate regardless of their native languages.
    d. organisms all have the same scientific name.

Answer: C Difficulty: I Section: 1 Objective: 2

27. Under the Linnaean system of classification, plants and animals are sorted into groups based on
    a. number and size.
    b. form and structure.
    c. form and size.
    d. number and structure.

Answer: B Difficulty: I Section: 1 Objective: 3

28. The largest division that a group of organisms can belong to is
    a. domain.
    b. class.
    c. genus.
    d. kingdom.

Answer: A Difficulty: I Section: 1 Objective: 3

29. Protista is an example of a
    a. kingdom.
    b. class.
    c. genus.
    d. species.

Answer: A Difficulty: I Section: 1 Objective: 3

30. Similar genera are grouped into a(n)
    a. phylum.
    b. class.
    c. family.
    d. order.

    Answer: C   Difficulty: I   Section: 1   Objective: 3

31. A species
    a. is a narrowly defined group of organisms.
    b. is a broadly defined group of organisms.
    c. has the same meaning as "population."
    d. None of the above

    Answer: A   Difficulty: I   Section: 1   Objective: 3

32. Each level of classification is based on
    a. specific characteristics.
    b. general characteristics.
    c. shared characteristics.
    d. All of the above

    Answer: C   Difficulty: I   Section: 2   Objective: 1

33. A biological species
    a. cannot interbreed within the natural population.
    b. is isolated reproductively from other species.
    c. can easily be differentiated from others based on appearance.
    d. produces infertile offspring.

    Answer: B   Difficulty: I   Section: 2   Objective: 2

34. A hybrid is produced from
    a. interbreeding between the same species.
    b. interbreeding between distantly related species.
    c. interbreeding between closely related species.
    d. crossing different plants.

    Answer: C   Difficulty: I   Section: 2   Objective: 2

35. Dogs and wolves are members of
    a. the same family.
    b. the same genus.
    c. different species.
    d. All of the above

    Answer: D   Difficulty: I   Section: 2   Objective: 2

36. The biological species concept is difficult to apply to
    a. sexually reproducing organisms.
    b. asexually reproducing organisms.
    c. organisms that produce pollen.
    d. organisms that live in groups.

    Answer: B   Difficulty: I   Section: 2   Objective: 2

37. interbreeding : hybrids
    a. water : soil
    b. mitosis : meiosis
    c. natural selection : change
    d. homologous : environment

    Answer: C   Difficulty: II   Section: 2   Objective: 2

38. Similar features evolved through convergent evolution are called
    a. analogous characters.
    b. homologous characters.
    c. environmental characters.
    d. genetic characters.

    Answer: A   Difficulty: I   Section: 2   Objective: 3

39. Convergent evolution produces similar features in different organisms as the result of
    a. similar environments.
    b. pressure by natural selection.
    c. sharing a common ancestor.
    d. Both (a) and (b)

    Answer: D   Difficulty: I   Section: 2   Objective: 3

40. Analogous structures
    a. have the same form in organisms.
    b. perform the same function in organisms.
    c. have the same structure in organisms.
    d. evolve from a common ancestor.

    Answer: B Difficulty: I Section: 2 Objective: 3

41. A model used by evolutionary biologists to represent the evolutionary history among species is called a
    a. phylogram.
    b. cladogram.
    c. histogram.
    d. parallelogram.

    Answer: B Difficulty: I Section: 2 Objective: 4

42. Derived characteristics are traits
    a. shared by all species.
    b. originated in a common ancestor.
    c. found in closely related species.
    d. found in distantly related species.

    Answer: C Difficulty: I Section: 2 Objective: 4

43. Evolutionary systematics emphasizes the importance of
    a. derived characteristics.
    b. unique characteristics.
    c. shared characteristics.
    d. compared characteristics.

    Answer: B Difficulty: I Section: 2 Objective: 4

COMPLETION

44. Aristotle grouped plants and animals according to their ____________________ similarities.

    Answer: structural Difficulty: I Section: 1 Objective: 1

45. The current system used for naming organisms was developed by ____________________.

    Answer: Linnaeus Difficulty: I Section: 1 Objective: 1

46. The two-word system for naming organisms is called ____________________ ____________________.

    Answer: binomial nomenclature Difficulty: I Section: 1 Objective: 2

47. All names assigned to organisms under the Linnaean system are in the ____________________ language.

    Answer: Latin Difficulty: I Section: 1 Objective: 2

48. The unique two-word name for a species is its ____________________ name.

    Answer: scientific Difficulty: I Section: 1 Objective: 2

49. The scientific name of an organism gives biologists a common way of ____________________ regardless of their native languages.

    Answer: communicating Difficulty: I Section: 1 Objective: 2

50. All living things are grouped into one of three ____________________.

    Answer: domains Difficulty: I Section: 1 Objective: 2

51. There are ____________________ levels of classification.

    Answer: eight Difficulty: II Section: 1 Objective: 3

52. A kingdom contains many ____________________.

    Answer: phyla Difficulty: II Section: 1 Objective: 3

53. Classes with similar characteristics are assigned to a(n) ____________________.

    Answer: phylum Difficulty: I Section: 1 Objective: 3

54. Each level of classification is based on ____________________ shared by all the organisms it contains.
    Answer: characteristics Difficulty: I Section: 1 Objective: 3

55. Taxonomy is a ranked system of groups that increase in ____________________.
    Answer: inclusiveness Difficulty: II Section: 1 Objective: 3

56. *Homo habilis, Homo erectus,* and *Homo sapiens* all belong to the same ____________________.
    Answer: genus Difficulty: I Section: 1 Objective: 3

57. Traditionally, scientists have used differences in appearance and ____________________ to classify organisms.
    Answer: structure Difficulty: I Section: 2 Objective: 1

58. A(n) ____________________ species is a group of actually or potentially interbreeding natural populations that are reproductively isolated from other such groups.
    Answer: biological Difficulty: I Section: 2 Objective: 2

59. A(n) ____________________ is produced through breeding closely related species.
    Answer: hybrid Difficulty: I Section: 2 Objective: 2

60. The formation of a new volcano that divides a population into two populations may result in ____________________ isolation of the newly formed populations.
    Answer: reproductive Difficulty: I Section: 2 Objective: 2

61. Many species of plants, some mammals, and many fishes are able to form ____________________ hybrids with one another.
    Answer: fertile Difficulty: II Section: 2 Objective: 2

62. Making evolutionary connections based on similar ____________________ can be misleading because not all traits are inherited from a common ancestor.
    Answer: traits Difficulty: I Section: 2 Objective: 3

63. The type of evolution that results in similar characteristics found in different organisms as the result of selection within similar environments is called ____________________ evolution.
    Answer: convergent Difficulty: I Section: 2 Objective: 3

64. Analogous structures are found in ____________________ taxa as a result of similar environmental conditions.
    Answer: different Difficulty: I Section: 2 Objective: 3

65. Homologous structures are found in organisms that once shared a(n) ____________________ ancestor.
    Answer: common Difficulty: I Section: 2 Objective: 3

66. The evolutionary history of a species is called its ____________________.
    Answer: phylogeny Difficulty: I Section: 2 Objective: 3

67. A system of taxonomy that reconstructs phylogenies by inferring relationships based on similarities is called ____________________.
    Answer: cladistics Difficulty: I Section: 2 Objective: 4

68. A model developed by taxonomists that diagrams the sequence of evolution between species is called a(n) ____________________.
    Answer: cladogram Difficulty: I Section: 2 Objective: 4

69. Cladistics is used to determine the ____________________ in which different groups of organisms evolved.
    Answer: sequence or order Difficulty: I Section: 2 Objective: 4

70. Species that have unique characteristics or traits that arose after divergence from other species are said to have ________________ traits.
    Answer: derived Difficulty: I Section: 2 Objective: 4

71. Animals that appear early on a cladogram do not have all of the same ________________ traits as the animals that appear later on the cladogram.
    Answer: derived Difficulty: I Section: 2 Objective: 4

72. Unlike cladistics, evolutionary systematics places more ________________ on some traits than on others.
    Answer: importance Difficulty: I Section: 2 Objective: 4

73. Evolutionary systematics requires ________________ information than cladistics.
    Answer: more Difficulty: I Section: 2 Objective: 4

ESSAY

74. How did Linnaeus's system of naming organisms simplify and improve the task of naming organisms?
    Answer:
    The polynomial system that existed prior to the Linnaean system of naming organisms produced long and cumbersome names that biologists often changed. Linnaeus gave each organism a two-word Latin name, reducing unwieldy names to something more manageable. Because his system has been universally adopted, scientists around the world can discuss the same organism regardless of their native languages.
    Difficulty: II Section: 1 Objective: 1

75. Describe the levels of classification in order of increasing inclusiveness.
    Answer:
    The species level is the least inclusive and most specific of all levels containing only those within a species that cannot, generally, interbreed with any organisms outside their own species. All species with similar characteristics are contained within a genus. All genera with similar characteristics are contained within a family. All families with similar characteristics are contained within an order. All orders with similar characteristics are contained within a class. All classes with similar characteristics are contained within a phylum. All phyla with similar characteristics are contained within a kingdom. All kingdoms are included within three domains.
    Difficulty: III Section: 1 Objective: 3

76. What characteristics do biologists use to classify organisms?
    Answer:
    Biologists usually define species according to appearance and structure. The biological species concept defines species according to their sexual reproductive potential. Cladistics focuses on sets of unique characteristics, while evolutionary systematics places greater importance on specific characteristics.
    Difficulty: II Section: 2 Objective: 1

77. A new species of animal is identified in the tropical climate of the northwest coast of Africa. The animal looks similar to a species located in the temperate climate of the southeast coast of North America. How could scientists explain the similarities between the two species?
    Answer:
    Because the organisms have developed in different environments and appear to have similar characteristics, scientists may hypothesize that the species shared a common ancestor in the recent geologic past.
    Difficulty: II Section: 2 Objective: 3

78. What is the relationship between environment and analogous structures?

    Answer:

    Analogous structures are physical characteristics shared by organisms from vastly different taxa such as the wings of birds and the wings of insects. While these structures perform similar functions for both groups, these analogous structures do not exist in both groups due to common ancestral origin, but rather exist as a result of convergent evolution. Similar structural characteristics in taxa so distantly related that they may be considered unrelated arise as a result of selective pressure under similar environmental conditions.

    Difficulty: II   Section: 2   Objective: 3

79. Compare and contrast cladistics with evolutionary systematics.

    Answer:

    Both cladistics and evolutionary systematics provide taxonomists with models that may be used to determine phylogenetic relationships between taxonomic groups, particularly species. Cladistics provides evolutionary biologists with an objective tool for organizing taxonomic groups as it weighs all characteristics equally. Evolutionary systematics applies a more subjective view as this approach assigns weights to characteristics based on the importance of the each characteristic to the survival of the organism.

    Difficulty: III   Section: 2   Objective: 4

9997265920 1 2 3 4 5 6